RESOURCE BOOK FOR THE TEACHING OF ASTRONOMY

AND INSTRUCTOR'S GUIDE FOR IVAN KING'S TEXT THE UNIVERSE UNFOLDING

ANDREW FRAKNOI

Cañada College

W. H. FREEMAN AND COMPANY
San Francisco

Cover Illustration: The Horsehead Nebula in Orion. Photo courtesy of the Kitt Peak National Observatory.

Printed in the United States of America

9 8 7 6 5 4 3 2 1

PREFACE
(BY AUTHOR OF INSTRUCTOR'S GUIDE)

"A delightful man, extremely modest for all his vast knowledge, candid as a child, delicately tactful and considerate. He responds with boyish gravity to all of one's questions, makes everything very clear, gives explanations which require little reflection on the part of the listener, and never loses his patience....He knows the history of all the heavenly bodies to the furthermost boundaries of the Galaxy and is more than willing to communicate to others all his wisdom."
--A contemporary account of William Herschel[1]

While perhaps few of us can hope to aspire to Herschel's skill as astronomer or instructor, I think most teachers of astronomy at least occasionally wish that they could do more to convey the excitement of modern astronomy to their students. Unfortunately, even those who have both the time and the inclination to do more than the minimum in their courses for nonscience majors are often not familiar with the many resources (other than a textbook) available to make this task easier.

Thus this resource book--intended to accompany Ivan R. King's text, *The Universe Unfolding*--is designed to provide astronomy instructors with an array of tools for more effective teaching. The heart of the book consists of what, to my knowledge, is the most extensive available list of references for teaching astronomy at the introductory level.

Specifically, resource material for each chapter is presented in the following sequence.

1. An *introductory paragraph* by Ivan King, explaining his intentions for the chapter. (This paragraph is in italics to emphasize that it is not by the author of the remainder of this resource book.)
2. Occasionally, *teaching notes*, pointing out difficult areas in the chapter, or sections which could be skipped.
3. *Summary of key points* in the chapter, together with a list of new terms.
4. *Topics for discussion and activities*, appropriate as mini-lectures, for small-group discussion, homework assignments, oral reports, and/or term papers.
5. *Reading materials for the student*, at approximately the same level as the text (i.e., very little or no math) and arranged by topics.
6. *Reading material for the instructor*: A representative sampling of review papers and books. Please note that this is *not* a listing

---
[1] Quoted in R. Thiel, *And There was Light* (Mentor paperback, 1957), p. 220.

of current research papers; rather it is a compilation of secondary sources, aimed primarily at the instructor who does not or cannot keep up with the *Astrophysical Journal*.

7. *Audio-visual aids*, including films, slides, and tapes, whenever available.
8. *Sample examination questions*, both multiple-choice and essay types.

Appendices summarize more general resources available to teachers and students of astronomy and give source addresses for materials listed throughout the manual.

The author of this resource book would welcome additions or corrections for each section, as well as comments on the effectiveness of the various suggestions. Please send these to Andrew Fraknoi, Cañada College, Redwood City, Calif. 94061.

At a time when growing public mistrust of science goes hand in hand with rising acceptance of astrology and other forms of mysticism, it becomes increasingly incumbent upon all scientists to contribute to public understanding of the scientific method and worldview. It is the author's hope that this resource book may be of some help in this vital endeavor.

--Andrew Fraknoi
San Francisco, November 1976

FOREWORD
(BY AUTHOR OF TEXT)

What should an elementary course in astronomy be? Every teacher has his own approach, and every student has his own expectations; yet beyond mere questions of style or personal preference, there exist some basic values and purposes. I have some strong opinions about elementary astronomy, and they are related in turn to even more basic opinions about the purpose of higher education. *The Universe Unfolding* is, in a sense, an expression of those opinions. The nature of the book will become clearer, I believe, if I elaborate on the point of view that made the book what it is.

A course in astronomy exemplifies one of the basic aspects of education: it offers learning, not training. Only the tiniest fraction of the students will make practical use of what they are taught in this course; the aim instead is to give them something more valuable but less tangible, something we call learning.

The learning experience in elementary astronomy has two sides: first, a student should appreciate what the Universe is and what place Man has in it; and second, he should see and experience some of the scientific reasoning that has drawn this picture. Stop for a moment, and think about these two objectives. Their full impact is overwhelming. If a course in astronomy were truly to bring a student to either of these goals--either to appreciate the role of Man in the Universe or to understand the workings of science--it would be one of the greatest experiences of his education. Even though in the real world, in real schools, we succeed in taking only small steps, we should never forget the goals toward which we strive.

In writing *The Universe Unfolding*, I have tried to keep these goals as firmly in mind as possible, and they determine the content and the tone of the book. Thus I have avoided cluttering the text with useless facts. Everything that is presented is included for a purpose; the student is never told that he has to learn something just "because it is there." A factual picture of the Universe is indeed built up, but only in the context of trying to understand it. This theme gives the text its flow and its continuity, which I have declined to break with frequent section and subsection divisions. Each chapter is meant to be read as a continuous narrative.

The theme of the text is understanding, which here means two things (as I have emphasized in the Preface to the book). Each of them is centered on an idea that the student should not just think but should feel. The first of these ideas is that the Universe is a rational structure--that things work as they do because they follow a set of basic rules that are uniform and understandable. The other important idea, the other sense of the word "understanding," refers to our own comprehension, and specifically to the ways in which we come to comprehend the Universe. Not only does the Universe itself make sense, but our own process of studying it makes clear and simple sense.

The logic of the Universe, and the logic of the scientist. These are the two central ideas toward which we should lead our nonscience students. For this, astronomy is a magnificent vehicle; this is why our students have chosen our subject rather than some other science. But in exciting our students about pulsars, black holes, and the expanding Universe, we should not lose sight of the fact that what we really offer them is an experience in logical thinking and its results.

This is an experience that many of them have not encountered elsewhere in their higher education. In too much of their study they have replaced reasoning by memorization; and where reasoning remains, it is all too often only a kind of uneasy intuition.

In saying this I do not mean to deprecate the teaching of our colleagues in other fields; I would emphasize instead the positive role that a course in astronomy can play. It is in this tone that I have written my book. Show the student how the Universe works, why it works that way, and how we know this. Such an approach need not make an elementary course any harder; in fact, it should make it easier, if these arguments carry any weight at all. The facts should be easier to remember, if they hang together and make sense.

The book is aimed at explaining astronomy to nonscientists, but I do not believe that the needs of the prospective science major are slighted. His aims are exactly to understand the Universe and to see how we find out about it. In approaching each area of astronomy, I have had to ask what the real questions are, what we really understand about them, and how we know it; and I hope that a student on any level can profit from the resulting presentation.

The one serious concession that I have made is the omission of mathematics. It has been my experience that the great majority of non-science students are so uncomfortable with mathematical manipulation that the use even of simple algebra is a serious barrier to their understanding. When I have required students to use simple mathematics, they have concentrated so heavily on the mathematical manipulations that they lose the astronomy. It is a sad retreat, but I do not believe that a course in elementary astronomy is the place to cure the mathematical illiteracy of the American college student.

On the other hand, I refuse to compromise on the subject of language. I have told the story of the Universe as I think it should be told, in the language that seems appropriate to it. Language is at the heart of everything that our students will ever learn, and expressive language should live in all their courses, not just in their English course. Moreover, astronomy is not just a list of facts or a table of numbers; it is a structure, an understanding, and a learning; it is an attitude and an appreciation. I have written the book to convey this spirit too.

The organization of the book is conventional; the Universe unfolds from the Earth outward. I think that it is important to start the student with everyday examples, in everyday surroundings, and to bring him gradually to the abstractions of stellar and extragalactic astronomy. The chapters are intended to be read in order, but an instructor who wants to give a short course that skips the Solar System can omit Chapters 3-11. Alternatively, a short survey of astronomy can skip Chapters 4 (Time, Place, and the Sky), 8 (The Copernican Revolution), 14 (Binary Stars), 15 (Variable Stars), 17 (The Sun), or any subset of these. A few things will then need to be filled in, but I have tried to provide an effective index to help the instructor or the student find them.

It gives me much satisfaction to place *The Universe Unfolding* in the hands of students, because in twenty years of teaching I have never found a text that satisfied the aims set forth above. I look forward to using it, and to doing classroom teaching instead of lecturing. I hope that you will enjoy it too.

--Ivan R. King
October 1976

TABLE OF CONTENTS

(Each chapter has the following sub-divisions:
A. Note by Author of Text
B. Teaching Notes (not in every chapter)
C. Summary of Key Points
D. Basic Terms Introduced in this Chapter
E. Topics for Discussion and Class Activities
F. Reading Materials for the Student
G. Reading Materials for the Instructor
H. Audio-Visual Aids
I. Sample Exam Questions: Multiple Choice (with answers)
J. Sample Exam Questions: Essays)

CHAPTER 1
THE SETTING

NOTE BY AUTHOR OF TEXT

I have always considered it important to begin a course in astronomy by orienting the student, in several separate ways. First, the sooner that he gets the distance hierarchy firmly in mind, the better. (Note that I have fixed firmly on the parsec as the unit of distance. There is no justification for the absurd practice of talking down to the public by translating everything into light years. Surely an honest textbook can use honest units.) The other part of the orientation is to tell the student what astronomy is about. The last two sections of the chapter do this; they set the tone of the book and state its themes, which will be developed throughout the remainder of the book.

--I.R.K.

TEACHING NOTES

The point about the immensity of cosmic distances cannot be overemphasized early in the course. Most students have thought of the stars and planets only as points of light on the dome of the sky. In particular, students need to be confronted with the fantastic changes of scale necessary in going from distances within the Solar System to those in the Galaxy--and then again, outside the Galaxy. See the first two of the "Topics for Discussion" for some suggestions.

SUMMARY OF KEY POINTS

Modern Astronomy

1. Stars are the fundamental building blocks of the Universe.
2. Since 1900 astronomical research has changed considerably:
    a. emphasis on stars and galaxies, rather than the Solar System;
    b. tremendous improvement in telescopes, as well as photographic and photoelectric techniques;
    c. use of detectors in the nonvisible regions of the electromagnetic spectrum.
3. Today astronomers are mostly applied physicists--questions shift from "where" and "what" to "why" and "how."
4. A fundamental assumption: The laws of physics are the same everywhere in the Universe.

Hierarchy of Distances

1. Basic unit used by astronomers is the *parsec*.
2. The distances to other stars are enormously greater than the distances within our Solar System. (This point cannot be overemphasized early in the course, because most students have thought all their lives only of points of light on the dome of the sky. See Topic #1 under "Topics for Discussion," below.)
3. Distances:
   a. between stars--a few parsecs;
   b. from one end of a galaxy to the other--tens of thousands of parsecs;
   c. between galaxies--a million parsecs;
   d. the observable Universe--billions of parsecs.

TOPICS FOR DISCUSSION AND CLASS ACTIVITIES

1. Almost every teacher or lecturer on astronomy likes to start a talk with some analogy concerning cosmic distances (like the marble-and-grain-of-sand comparison of our text). But it can be instructive for the student to make up such an analogy of his or her own, particularly since this is a good way to begin to think about some of the unthinkably large numbers involved. (This is also a good exercise for explaining or reviewing scientific notation--with which, lamentably, not every high school student these days is familiar.) Here, for your conveneience, is a *table of distances* in kilometers.

   Earth diameter = 1.3×10^4 km
   Distance to the Moon = 3.8×10^5 km
   Sun diameter = 1.4×10^6 km
   Distance to the Sun = 1.5×10^8 km
   Distance from the Sun to Jupiter = 7.8×10^8 km
   Distance from the Sun to Pluto = 5.9×10^9 km
   Distance to nearest star = 4×10^{13} km
   Distance to the center of the Galaxy = 3×10^{17} km
   Galaxy diameter = 10^{18} km
   Distance to Andromeda galaxy = 2×10^{19} km
   Distance to 3C123 (most distant known normal galaxy) = 8×10^{22} km

   Some ways to proceed might be to pick two favorite local cities to represent the Sun and Proxima Centauri, or to take the U.S. to represent the Galaxy.

   After working through such an exercise, students should be ready to appreciate the following quote from *The Sot Weed Factor*, a novel by John Barth (Bantam paperback, 1967):

   > ". . . for the first time in his life he saw the night sky. The stars were no longer points on a black hemisphere that hung like a sheltering roof above his head; the relationship between them he saw now in 3 dimensions, of which the one most deeply felt was depth. . . what struck him now was that some were nearer, others farther out, and others unimaginably remote. Viewed in this manner, the constellations lost their sense entirely; their spurious character revealed itself, . . . and Ebenezer felt bereft of orientation. He could no longer think of up and down: the stars

were simply *out there*, as well below him as above, and the wind appeared to howl not from the Bay, but from the firmament itself, the endless corridors of space."

2. Another good way to get the students to think about the cosmic scale of things is to show the film *Powers of Ten*--or, if necessary, its silent version, *Cosmic Zoom*, which is not as good (see "Audio-Visual Aids," below), and/or to have them look at the book that inspired the films: *Cosmic View: The Universe in 40 Jumps*, by Kees Boeke (John Day, 1957). Both the book and the films move away from the Earth in jumps of one power of ten of increasing distance, and then move into the microscopic realm in the same way.

3. Many laypersons still think of astronomical research the way people did in the days of Kepler and Halley: comets, shooting stars, naming new objects, etc. Thus it is particularly important to stress at the outset that modern astronomy puts far greater emphasis on stars and collections of stars (as well as interstellar matter) than on local phenomena in our Solar System. To emphasize this, it can be valuable to contrast astronomical work during earlier centuries with the work today, for example by showing slides of the best instruments, observatories, publications, results, etc. If a university library has some old astronomy textbooks, reading selected outdated passages can be instructive.

4. A more difficult project along the same lines would be to contrast a day in the life of the average astronomer in the sixteenth, eighteenth, and twentieth centuries. For example, students might select Tycho Brahe, William Herschel, and Edwin Hubble. Or select a living astronomer for the twentieth-century choice--perhaps even someone close by, with whom a representative of the class could talk.

   Here are some references with which students might begin.

   a. *Tycho Brahe.*
      Christianson, J. "The Celestial Palace of Tycho Brahe," *Scientific American*, February 1961.
      Dingle, H. "Tycho Brahe," in *Astronomy*, ed. S. Rapport and H. Wright (Washington Square Press paperback, 1964).
      Thiel, R. "An Observer and a Prophet," in *And There was Light* (Mentor paperback, 1957).
   b. *William Herschel.*
      Hoskin, M. *William Herschel and the Construction of the Universe* (Norton, 1964).
      Hoskin, M. "Herschel: The Depths of Space and Time," in *The Mind of the Scientist* (Taplinger, 1971).
      Kopal, Z. "The Space Widens," in *Widening Horizons* (Taplinger, 1970).
      Thiel, R. "Herschel's Plunge into Space," in *And There Was Light* (Mentor paperback, 1957).
   c. *Edwin Hubble.*
      Goldsmith, D. "Edwin Hubble and the Universe Outside Our Galaxy," in *The Heritage of Copernicus*, ed. J. Neyman (MIT Press, 1974).

Hubble, E. *The Realm of the Nebulae* (Dover paperback, 1936). [Classic summary of his work; see especially Chapter 1.]

Time magazine, cover story, issue of February 9, 1948.

d. *Living astronomers.*

"Jesse Greenstein at the 200-inch," *Horizon*, Summer 1967 issue.

Articles on Margaret and Geoffrey Burbidge in *Smithsonian*, January 1967 issue; *Science Year 1969* (science supplement to *World Book Encyclopedia*); and *Mercury*, July/August 1975 issue.

Mercury magazine has an on-going series of interviews with prominent astronomers under the title "Observing Astronomers."

5. The text mentions that "this year's crop of new astronomers could have staffed the whole astronomical world of a hundred years ago." Discuss (or have students research) what problems this sharp increase in the number of astronomers is causing in the astronomical community.

Some references:

Astronomy Manpower Committee of the Committee on Science and Public Policy. *Employment Problems in Astronomy*. (National Academy of Sciences, Washington, D.C. 20418).

Bok, B., and D. Goldsmith. "Present Employment Trends in Astronomy," *Mercury*, July/August 1973.

National Academy of Sciences. *Astronomy and Astrophysics for the 1970's*, vol. 1 (National Academy of Sciences, 1972). [A general report on the state of the astronomical world.]

Science News, March 15, 1975, p. 206.

6. It is often helpful to start a new astronomy class by taking a little survey of the students, to have them think about how much (or, more likely, how little) they know about the Universe and astronomy. Then it might be good to discuss where the failure lies: in the students, in the American educational system, in the astronomical community, in the media, etc.

Among questions that have successfully sparked student discussion in my classes are the following.

a. Can you name a living astronomer? (How many can you name?) Contrast this with the number of sports personalities, novelists, composers, painters, architects, etc.
b. Could you explain why the sky is blue?
c. Could you explain why the Sun shines and keeps on shining?
d. How old is the Solar System?

7. The text emphasizes the assumption that the laws of physics (the rules of the game, if you will) are the same everywhere in the Universe. Discuss the problems (philosophical and practical) that would arise if this were not so. Have students discuss evidence that this assumption holds, for example, on the Moon or in a distant galaxy. (This question could well be taken up again after discussing spectral lines in stars.)

Some physicists have speculated about subsequent cycles of an oscillating Universe in which the laws of physics may be different. (See, e.g., J. Wheeler, "The Universe as Home for Man," *American Scientist*, Nov.-Dec. 1974.)

8. Even more fundamental (and abstract) is the assumption with which this chapter closes--that "the workings of the Universe can be understood logically." Many students these days have serious doubts about this assumption and subscribe to various forms of mysticism, some of which explicitly deny the rationality of the Universe.

   Some instructors may like to discuss this question early in the course (it can lead to lively debate) and without proselytizing to explain the worldview of modern science more explicitly.

9. Discussions such as the one just suggested frequently lead to questions about astrology, particularly when some of the growing legion of true believers are members of the class.

   Here are some useful references for an objective look at astrology.

Abell, G. "Astrology: Its Principles and Relation and Nonrelation to Science," *Science Teacher*, December 1974.

Barth, J., and J. Bennett. "Astrology and Modern Science Revisited," in *Leonardo*, vol. 7, p. 235 (1974).

Bok, B., and L. Jerome. *Objections to Astrology* (Prometheus Books paperback, 1975). [Reprint of a special section in the Sept.-Oct. 1975 issue of *Humanist* magazine, beginning with a statement signed by 186 leading scientists, raising general objections to astrology, and then featuring two critical articles by astronomer Bart Bok and by science writer Lawrence Jerome, looking at specific problems. The statement alone is reprinted in *Science News*, September 13, 1975.]

Bok, B., and M. Mayall. "Scientists Look at Astrology," *Scientific Monthly*, vol. 52, p. 233 (1941).

Fraknoi, A. "Debunking Pseudo-science," *Mercury*, Nov.-Dec. 1975. [Includes a general bibliography on all sorts of pseudo-science topics.]

Frazier, K. "Science and the Parascience Cults," *Science News*, May 29, 1976. [Chronicles some attempts to challenge pseudo-science.]

Jerome, L. "Astrology and Modern Science: A Critical Analysis," *Leonardo*, vol. 6, p. 121 (1973).

Jerome, L. Tape of an interview with Jerome, coauthor of the statement against astrology (Astronomical Society of the Pacific, radio program cassette ER-141).

Reaves, G. "The Rise and Fall of Astrology," in *A.S.P. Leaflet* #480 (June 1969). [These leaflets were published by the Astronomical Society of the Pacific until they were replaced by *Mercury* magazine; remaining copies are available for 10¢ and a stamped self-addressed envelope. See address in Appendix 2.]

Reis, R. "Observing Astronomers: George Abell on Astrology and Astronomy," *Mercury*, March/April 1976.

Our text discusses astrology in Chapter 7. In the absence of questions from students, some instructors may prefer to postpone these kinds of discussion until then (or, perhaps, forever).

10. In the late 1950s, the British scientist/novelist C. P. Snow proposed that the world was slowly dividing into camps, scientific and humanistic, and that there was a growing gap of communication between these "two cultures." He felt that this posed a major problem in our increasingly technological society. Has this division (and any problems it may cause) increased or decreased since that time? If students can be encouraged at this point to bring in their own feelings (fears, prejudices, apathy, etc.) toward science, quite a meaningful discussion can result, which might help to "clear the air" early in the course.

References:

Schroeer, D. *Science and Its Fifth Dimension: Society* (Addison-Wesley paperback, 1972). [Chapter 1 is a brief summary of the Snow controversy.]

Snow, C. *The Two Cultures and a Second Look* (Mentor paperback, 1963).

11. The previous discussion might well lead the class to ask if there can be profitable interplay between the scientific and humanistic cultures--between astronomy and poetry, for example. Here are two points of view on this subject to begin discussion.

First, a poem by Walt Whitman:

When I heard the learn'd astronomer,
When the proofs, the figures, were ranged in columns before me,
When I was shown the charts and diagrams, to add, divide and measure them,
When I sitting heard the astronomer when he lectured with much applause in the lecture-room,
How soon unaccountable I became tired and sick,
Till rising and gliding out I wander'd off by myself,
In the mystical moist night air, and from time to time,
Look'd up in perfect silence at the stars.

Second, a quote from the *Feynman Lectures on Physics* (Addison-Wesley, 1963) by Nobel-prize-winning physicist Richard Feynman:

"It does no harm to the mystery to know a little about it. For far more marvelous is the truth than any artists of the past imagined! Why do the poets of the present not speak of it? What men are poets who can speak of Jupiter if he were like a man, but if he is an immense spinning sphere of methane and ammonia must be silent?"

Other references on astronomy and poetry:

Fraknoi, A., and A. Friedman. "Images of the Universe," *Mercury*, March/April 1975.

Meadows, A. *The High Firmament* (Leicester Univ. Press, 1969).

Phillips, R., (ed.). *Moonstruck: An Anthology of Lunar Poetry* (Vanguard Press, 1974).

Wakowski, D. *Dancing on the Grave of a Son of a Bitch* (Black Sparrow Press, 1974). [See Section I.

"Astronomer Poems."]

In the same vein, some students may like to read C. Ronan, "Astronomy and Music," *Sky and Telescope*, September 1975.

NOTE

Because this chapter is introductory, no specific reading materials or examination questions have been included. Appendix 1 gives a list of general reference materials in astronomy with which students and teachers may wish to acquaint themselves.

AUDIO-VISUAL AIDS

Films

Powers of Ten (1968, color, 8 minutes) or a silent version, *Cosmic Zoom* (1970, color, 8 minutes), are brief, excellent "grand tours" of the physical universe. See Topic #2 under "Topics and Activities," above. *Powers of Ten* also includes a nice graphic demonstration of relativistic time dilation. (*Powers of Ten* was produced by Charles Eames for the Commission on College Physics; *Cosmic Zoom* was produced by the National Film Board of Canada and is distributed by McGraw-Hill films.)

Universe (1960, black and white, 28 minutes) is a less rapid tour of the cosmos. Though not completely up-to-date, it is superbly produced with a number of rather dramatic special effects. (Produced by the National Film Board of Canada; distributed by McGraw-Hill films.)

Why Man Creates (1969, color, 25 minutes)--especially episodes 1 and 6. An Academy-Award-winning film examining creativity, both artistic and scientific, from a number of unusual viewpoints. Uses animation and live photography to achieve some very clever effects. Not directly relevant to astronomy, but excellent for discussions of science and the humanities or the creative aspects of scientific research. (Pyramid Films.)

CHAPTER 2
LOOKING AT THE STARS

NOTE BY AUTHOR OF TEXT

The placement and the length of this chapter are compromises. Although an empirical approach to astronomy requires that methods of observation be introduced at an early stage, the intrusion of too much preliminary material can completely deaden students' interest long before they encounter the exciting problems of "real" astronomy. Thus I have kept this chapter as short as I could, covering only the actual radiation and the equipment with which we measure it. The physics of spectra, the Doppler effect, and all other physical laws and principles are brought in later, at points where they become relevant. Not only does this get the student into the meaty material quickly; it also emphasizes, at those later points, the intimacy of astronomy's relationship with physics.

--I.R.K.

TEACHING NOTES

Most students, as they enter an introductory course, will have thought only about visible-light observations, if they've thought about astronomy at all. Thus it is particularly important to stress early in the course to what degree modern astronomy involves working with all the wavelengths of the spectrum. It probably helps to make an effort to relate the less familiar regions of the spectrum to processes and technology that are part of the students' everyday experiences.

SUMMARY OF KEY POINTS

Electromagnetic Radiation

1. The spectrum (in order of increasing wavelength): gamma rays, X-rays, ultraviolet, visible light, infrared, radio.
2. The Earth's atmosphere is transparent only in the optical and radio "windows."

Telescopes

1. Main characteristic: "light"-gathering power, which is determined by the collecting area.
2. Auxiliary instrumentation:

a. photographic plate--accumulates photons;
b. filters or spectrograph--separate various wavelengths;
c. photoelectric cell and image tubes--electronic photoamplifiers.

3. Resolution: 70 λ's give a sharpness of 1°; better resolution requires a larger telescope or a shorter wavelength.

"Invisible" Astronomy

1. Progress required advances in technology.
2. Radio and selected infrared wavelengths are best explored, since they can be detected from the Earth's surface.
3. Short-wavelength studies must be done above the Earth's atmosphere.

Basic Terms Introduced in This Chapter

Electromagnetic radiation
Wavelength
Seeing

TOPICS FOR DISCUSSION AND CLASS ACTIVITIES

1. All the discussion of this chapter takes place in terms of wavelengths, although the relationship between wavelength and energy is hinted at on p. 18. Some instructors may want to make the relationship quite precise at this point, and also to talk about the concept of frequency.

2. On p. 16, the relationship between the transparency of the atmosphere and the wavelength sensitivity of the human eye is briefly mentioned. Students may enjoy pursuing this subject just a bit further. The atmosphere can't be the only determining factor, or else some organisms ought to have "radio eyes." After class discussion of resolution and collecting area, students should be able to come up with the reason that radio-sensitive organs are impractical.

   The other determining factor in this problem is, of course, the spectrum of the original source of the radiation by which we see--in this case, the Sun. Thus the class may want to return to this question after studying black-body spectra and solar radiations.

3. Suppose we ignore the practical considerations cited above, and ask what the sky would actually look like if our senses were sensitive to radio waves. What familiar objects would be "invisible"; what new objects would we "see"? This question might be broached at this point in the course and then postponed for further discussion until students have learned more about radio astronomy. One answer to this problem can be found in a map on p. 297 of *Radio Astronomy* by J. Kraus (McGraw-Hill, 1966).

4. In stressing what a difference telescopes make in terms of our ability to understand more about the Universe, it is instructive to relate the story of the first recorded use of a telescope by Galileo in the first decade of the 1600s. His initial three discoveries--the Galilean moons of Jupiter, the phases of Venus,

and sunspots--were all revolutionary challenges to the cosmology of his time.

5. Although the last two decades have seen a dramatic upswing in the exploration of the nonvisible regions of the electromagnetic spectrum, until recently one band (the so-called extreme ultraviolet) remained unexplored. However, during the summer of 1975, the Apollo-Soyuz mission involved the use of the first extreme-ultraviolet telescope for non-Solar-System astronomy. See:

Eberhart, J. "Astrodiplomacy in Orbit," *Science News*, July 5, 1975.
Margon, B., and S. Bowyer. "Extreme Ultra-Violet Astronomy from Apollo-Soyuz," *Sky and Telescope*, July 1975.

It might be an interesting exercise to have students ignore the above two articles at first, and try to find out about the extreme-ultraviolet experiment through the non-science-oriented media. It may be an insuperable task; N.A.S.A. apparently decided to downplay the science and play up the politics for the Apollo-Soyuz mission--a move that, according to some observers, gravely misjudged the public mood and lent a great air of triviality and financial waste to the mission. Would the students in the class agree?

6. The chapter mentions that the Earth's ionosphere makes long-distance radio communications possible. TV waves, on the other hand, are not reflected by any region of the atmosphere, and they travel through. The problem of long-distance TV transmission was solved through the use of communications satellites, which can thus be regarded as an artificial ionosphere. (By the way, the idea of communications satellites apparently first occurred to science-fiction writer Arthur Clarke.) This might be a nice topic for a paper for students interested in space technology.

7. If TV waves are not stopped by our atmosphere, then they must continue traveling outward. At the speed of light, our earliest TV transmissions must now be about 8 parsecs (25 light years) away. Some writers have speculated that an advanced civilization may have the technology to detect the extremely weak signals. Is someone out there learning about the Earth by watching early episodes of *I Love Lucy*? What would they conclude about the existence of intelligent life here? (Also see Chapter 11 of the text.)

8. The chapter discusses how astronomical "seeing" depends on various natural factors in the Earth's atmosphere. Recently some man-made factors have been added to the problems of astronomers. The most annoying problem is the "light pollution" caused by our ever-expanding cities and suburban areas, particularly in California. There may come a time when some of our biggest telescopes will have to be relocated at great expense, to escape the brightening skies of San Jose and San Diego. Yet, in many cases, a simple top-side reflector placed on all lights could eliminate a large fraction of the problem. Such reflectors would keep light out of the sky and down on the ground where it is needed. See:

Cuffey, J. "The Spectrum of a City," *Sky and Telescope*, September 1975.
Hoag, A. "Observatories and City Lights," *Mercury*, Sept./Oct. 1972.
Nelson, B. "Can We Save What's Left of the Dark?" *Mercury*, Jan./Feb. 1973.
Riegel, K. "Light Pollution," *Science*, March 30, 1973.
Thomsen, D. "Light Pollution," *Science News*, December 15, 1973.

9. The development of radio and infrared astronomy makes interesting case studies for examining the interaction of science and technology. Here are some references to get students started in the history of each field.

 a. *Radio astronomy.*
 Hey, J. *The Evolution of Radio Astronomy* (Neale Watson, 1973).
 Pfeiffer, J. *The Changing Universe* (Random House, 1956).
 Verschuur, G. *The Invisible Universe* (Springer Verlag paperback, 1974).
 Westerhout, G. "The Early History of Radio Astronomy," *Annals of the New York Academy of Science*, vol. 198 (August 25, 1972).
 b. *Infrared astronomy.*
 Allen, D. *Infrared, The New Astronomy* (Halsted, 1976).
 Barr, E. "Historical Survey of the Early Development of the Infra-red Spectral Region," *American Journal of Physics*, January 1960.
 Murray, B., and J. Westphal. "Infrared Astronomy," *Scientific American*, August 1965.
 Ronan, C. *Invisible Astronomy* (Lippincott, 1972).

READING MATERIALS FOR THE STUDENT

Telescopes (Optical)

Philip, A. "A Visit to the Soviet Union's 6-Meter Reflector," *Sky and Telescope*, May 1974.
"Photographic Report from Kitt Peak," *Sky and Telescope*, January 1973.
Thomsen, D. "The Many-mirrored Telescope of the Future," *Science News*, August 18/25, 1973.
Wampler, J. "The 153" Anglo-Australian Telescope," *Sky and Telescope*, October 1975.

Telescopes (Radio)

"Arecibo's Great Radio Telescope Upgraded," *Sky and Telescope*, March 1975.
Findlay, J. "The National Radio Astronomy Observatory," *Sky and Telescope*, December 1974.
Frazier, K. "Future World Center of Radio Astronomy" [about the Very Large Array], *Science News*, August 24/31, 1974.
Hachenberg, O. "The New Bonn 100-Meter Radio Telescope," *Sky and Telescope*, December 1970.
Heeschen, D. "The Very Large Array," *Sky and Telescope*, June 1975.
Shawcross, W. "Arecibo Observatory Today," *Sky and Telescope*, May 1972.

Verschuur, G. *The Invisible Universe* (Springer Verlag paperback, 1974).

<u>Instrumentation, Resolution, and Seeing</u>

Bracewell, R. "Stanford's High-resolution Radio Interferometer," *Sky and Telescope*, July 1971.
Kellermann, K. "Intercontinental Radio Astronomy," *Scientific American*, February 1972.
Smith, M. "AURA's Fabry-Perot Interferometer," *Sky and Telescope*, April 1973.
Thomsen, D. "Photographing the Stars Electronically," *Science News*, May 6, 1972.
Thomsen, D. "Image Tubes," *Science News*, November 17, 1973.
Thomsen, D. "A Sharper View of the Stars," *Science News*, August 24, 1974.
Thomsen, D. "Stars of Sky and Screen," *Science News*, August 23/30, 1975.
Thomsen, D. "Toward the Diffraction Limit," *Science News*, August 23/30, 1975.
Weymann, R. "Extending the Visible Frontier: New Tools of the Optical Astronomer," *Mercury*, Sept./Oct., 1975.
Young, A. "Seeing and Scintillation," *Sky and Telescope*, September 1971.

<u>Space Astronomy</u>

Burbidge, M., and G. Field. "The Space Telescope," *Mercury*, July/August 1976.
Kratage, M., and D. Wood. "Space Astronomy," *Astronomical Society of the Pacific Leaflet* #489 (March 1970).
Sagan, C. "Experimental Astrophysics," *Mercury*, March/April 1975.
Science News. Special issue on Skylab. May 5, 1973.
Stehling, K. "Balloon Astronomy," *Smithsonian*, June 1971.

<u>Invisible Astronomy (Introductory Materials)</u>

Edson, L. "X-rays from the Sky," in *Science Year 1974* (World Book Encyclopedia).
Goldberg, L. "Ultra-violet Astronomy," *Scientific American*, June 1969.
Hjellming, R. "Radio Astronomy," in *McGraw-Hill Yearbook of Science and Technology*, 1974.
Kraushaar, W., and G. Clark. "Gamma Ray Astronomy," *Scientific American*, May 1962.
Neugebauer, G., and E. Becklin. "The Brightest Infra-red Sources," *Scientific American*, April 1973. Reprinted in *New Frontiers in Astronomy*, ed. O. Gingerich, W. H. Freeman and Co., 1975.
Ronan, C. *Invisible Astronomy* (Lippincott, 1972).
Thomsen, D. "The Mystery of the Cosmic Gamma-ray Zaps," *Science News*, June 1, 1974.

READING MATERIALS FOR THE INSTRUCTOR

Barlow, B. *The Astronomical Telescope* (Springer-Verlag, 1975).
Code, A. "New Generation Optical Telescope Systems," in *Annual Reviews of Astronomy and Astrophysics*, vol. 11 (1973).
Dennison, E. "Electronic Optical Astronomy: Philosophy and Practice,"

Science, October 15, 1971.
Kuiper, G., and B. Middlehurst. *Telescopes* (Univ. of Chicago Press, 1960).
LaLonde, L. "The Upgraded Arecibo Observatory," *Science*, October 18, 1974.
Livingston, W. "Image Tube Systems," in *Annual Reviews of Astronomy and Astrophysics*, vol. 11 (1973).
Metz, W. "TV Cameras are Replacing Photographic Plates," *Science*, March 31, 1972.
Millikan, A. "Image Detection at the Telescope," *American Scientist*, May/June 1974.
Ryle, M. "Radio Telescopes of Large Resolving Power" [Nobel Prize lecture], *Science*, June 13, 1975.
Steinberg, J., and J. Lequeux. *Radio Astronomy* (McGraw-Hill, 1963).
Swenson, G., and K. Kellermann. "An Intercontinental Array: A Next Generation Radio Telescope," *Science*, June 27, 1975.

AUDIO-VISUAL AIDS

Films

The Astronomer (1960, color, 16 min). A general introductory film, focusing (pun intended) on the 200-inch telescope and its auxiliary instruments, with a brief mention of a radio telescope. (Distributed by the International Film Board.)

Electromagnetic Waves (1957, black and white, 33 min). A physics film featuring G. Volga of MIT doing experiments to show the wave nature and essential similarity of light waves, X-rays, and radio waves. (A PSSC film, distributed by Modern Learning Aids.)

Four Telescope Arrangements (1966, black and white, 4 min). A silent film strip showing various arrangements of the elements in a reflector telescope, using a ray-streak board. (A BBC School Television Loop, distributed by Gateway Educational Films, Ltd.)

Frontiers of Space (1962, color, 11 min). A brief (and rather superficial) survey of light and telescopes. (Distributed by Encyclopaedia Brittanica Films.)

Journey into Light (color, 25 min). The story of Kitt Peak, the national astronomical observatory, operated jointly by several universities and funded by the federal government. (Available from Kitt Peak National Observatory, Photo Dept., P.O. Box 4130, Tucson, Arizona 85717.) See also their film *Skies of the Andes* (1962, color, 30 min) about the building of an observatory in the mountains of Chile.

Radiant Energy and the Electromagnetic Spectrum (1969, color, 11 min). A short and elementary summary of the discovery, sources, applications, and detection of the various bands of the spectrum. (Produced and distributed by Coronet Instructional Films.)

Radio Sky (1966, color, 28 min). A review of radio astronomy. (British film, available from Associated Electrical Industries, 33 Grosvenor Place, London S.W. 1, England.)

The Realm of the Galaxies (1960, color, 19 min). A nicely done little film featuring Allan Sandage, particularly noteworthy for its fine description of a night at the 200-inch telescope. (Part of the Horizons of Science series, distributed by the Educational Testing Service.)

Universe from Palomar (1968, black and white, 30 min). History of the

200-inch telescope (and its predecessors), and a description of its auxiliary equipment as well as some research projects. (An N.E.T. film, distributed by Indiana University.)

SAMPLE EXAM QUESTIONS (MULTIPLE-CHOICE)

1. Which of the following is *not* a type of electromagnetic radiation?
   a. light
   b. radio waves
   c. X-rays
   d. gravity waves
   e. infrared waves
2. Which of the following types of radiation is usually produced in nuclear reactions?
   a. gamma rays
   b. X-rays
   c. infrared
   d. ultraviolet
   e. radio waves
3. The main function of a telescope is generally to--
   a. magnify star images
   b. separate light into its spectrum
   c. help us see radiation that cannot get through the Earth's atmosphere
   d. collect more radiation from faint objects
   e. make more jobs for astronomers
4. The instrument that combines the accuracy and sensitivity of the photocell with the wide field capability of the photograph is the--
   a. telescope
   b. spectroscope
   c. image tube
   d. camera
   e. X-ray detector
5. Ultimately, the ability of a *large* telescope on a mountain top to *magnify* is limited by--
   a. the type of focus it has
   b. the photographic film we use
   c. the diameter of the collecting lens
   d. the Earth's atmosphere
   e. the time of film exposure
6. When the light from the primary mirror of a telescope is reflected by a small secondary mirror through a hole in the primary, astronomers call this--
   a. Cassegrain focus
   b. prime focus
   c. Newtonian focus
   d. coudé focus
   e. out-of-focus
7. Where would be the best place to locate a telescope for observing celestial X-rays?
   a. near the ocean
   b. near a hospital
   c. in a desert

d. on a mountain top
e. aboard a space station

8. A big radio telescope can be made very much larger than a big visible-light telescope. This is because the radio telescope--
   a. cannot be used to detect waves from very far away, so it is only used for local observations
   b. is built out of a set of special plastic-coated glass lenses that are cheap to grind
   c. doesn't get any interference at all from Earth sources
   d. detects waves so large that the reflecting surface can be quite rough and irregular (in fact, can even have holes in it)

Answers to Multiple-choice Questions

1. d 2. a 3. d 4. c 5. d 6. a 7. e 8. d

SAMPLE EXAM QUESTIONS (ESSAY)

1. Name and describe briefly the various regions of the electromagnetic spectrum, in order of increasing wavelength.
2. In which bands of the electromagnetic spectrum can celestial phenomena be observed from the Earth's surface? What interferes in the other cases?
3. Compare and contrast the uses of the following devices "attached" to a large telescope: (a) the human eye; (b) photographic plate; (c) spectrograph; (d) image tube. What are the advantages and limitations of each device?

CHAPTER 3
THE EARTH

NOTE BY AUTHOR OF TEXT

The chapter about the Earth aims in part at easing the transition to the Universe; we have our feet on the ground and our heads in the sky. New physical ideas--mass, density, atomic motions, etc.--are introduced in a familiar rather than an exotic context, and most of the facts and figures have some relation to everyday life. (Note that the distances are quoted in miles as well as in kilometers, to emphasize the familiar.) Yet the orientation is always toward studying the Earth as a planet, and particularly as a body that exemplifies the physical processes that have made it what it is.

--*I.R.K.*

TEACHING NOTES

Although some of the basic concepts introduced in this chapter will be necessary later, the later sections could be skipped by instructors teaching a brief or modern-astronomy-oriented course. One note of caution: given today's grade schools, it cannot be assumed that every student knows about the seasons.

SUMMARY OF KEY POINTS

Physical Characteristics

1. Radius--6,371 km (first measured by Eratosthenes).
2. Mass--6×10^{27} gm.
3. Mean density--5.5 g/cm^3 (interior denser, exterior less dense).
4. Surface temperature--higher than equilibrium with solar radiation would produce, because of greenhouse effect.

Atmosphere

1. Escape velocity as compared to mean molecular velocity determines what remains and what escapes.
2. Hydrogen and helium can escape easily.
3. Temperature gradient is determined by--
   a. vertical mixing;
   b. input of solar energy.
4. Layers: troposphere, stratosphere, ionosphere, exosphere.
5. Magnetic field dominates outermost regions--produces Van Allen belts.

Surface

1. Plates supporting the continents and ocean floors slowly drift.
2. Processes molding the land:
   a. tectonic activity--two plates driven against each other build up mountains;
   b. erosion by water, wind, and ice.

Interior

1. Seismology, the study of the propagation of earthquake waves, can tell us about the density and structure of the interior.
2. Structure:
   a. crust--30 km;
   b. mantle--2,900 km;
   c. outer core--liquid;
   d. inner core--probably solid.

Composition

1. Most abundant elements in the Earth: oxygen, silicon, aluminum, and iron.
2. Most abundant elements in the Universe: hydrogen, helium, carbon, nitrogen, oxygen, neon, and silicon.
3. The lighter elements were mainly in gaseous form and were lost; the heavier ones remained as part of the Earth.

Motions

1. Rotation--results in day/night, centrifugal force, and Coriolis force.
2. Revolution--connected with seasonal changes.

Basic Terms Introduced in This Chapter

Centrifugal force
The law of gravitation
Density
Mass
Sidereal day
Coriolis force
Solstices and equinoxes
Temperature as a measure of the motion of molecules
Greenhouse effect

TOPICS FOR DISCUSSION AND CLASS ACTIVITIES

1. Suppose Mars were inhabited by intelligent, curious astronomers, equipped with the same equipment as their present-day Earth counterparts. Would they be able to tell that the Earth is inhabited? If yes, how? If not, how close would they have to be before it became apparent? This question is discussed in some detail in Section I of Carl Sagan's book, below. References:

   Ponnamperuma, C., and A. Cameron (eds.). *Interstellar Communication* (Houghton Mifflin paperback, 1974). [See Chapter 1.]
   Sagan, C. *Planetary Exploration* (Univ. of Oregon Books, 1970).

2. With the advent of the space age, we now have photographs and live descriptions of the Earth seen in its entirety. Several astronauts reported being profoundly moved seeing our planet as

a single, small blue-green ball in the distance. If the teacher is so inclined, slides of the Earth might be an occasion for a discussion of an astronomical perspective on our problems on this planet. Several essays along this line appear in Carl Sagan's highly readable book for the layperson, *Cosmic Connection* (Dell paperback, 1973); see review in *Mercury*, March/April 1975.

Slides from space of the Earth are available from N.A.S.A. and from several commercial distributors. N.A.S.A. also distributes a large number of films free to educational institutions. (See Appendix 2 for addresses.)

3. Another recent facet of the study of our planet is the use of artificial satellites for remote sensing. Among the most extensive of such studies have been the Skylab and Earth Resources Technology Satellite programs. Here are a few useful references for students interested in further investigating these programs.

Canby, T. "Skylab, Outpost on the Frontier of Space," *National Geographic*, October 1974.
Cromie, W. "Research in Orbit," in *Science Year 1975* (World Book Encyclopedia).
Maugh, T. "ERTS: A New Way of Viewing the Earth," *Science*, April 13, 1973.
Nash, E. "Cataloging the Whole Earth," *Science Year 1975* (World Book Encyclopedia).
Science News. Special issue on Skylab. May 5, 1973.

4. The text mentions some of the differences between the composition of the Earth and the general abundances of the elements in the Universe. It is an interesting question to ask how significant these differences were for the formation of life on Earth. (See, for example, E. Frieden, "The Chemical Elements of Life," *Scientific American*, July 1972.)

5. There is geological evidence that in past epochs the Earth's magnetic field has undergone reversals. As this occurs, there must be a period when the field is zero and the Van Allen belts spill their trapped particles onto the Earth. The suggestion has been made that during such periods the rate of mutations and therefore of evolution increases. It is interesting to speculate on the effect a complete absence of a magnetic field might have had on the development of life on Earth.

6. Looking at aurorae in greater detail often makes for a popular student project. See, for example:

Akasofu, S. "The Aurora," *Scientific American*, December 1965.
Elvey, C. "Aurora and Airglow," *Scientific American*, September 1955.

7. Suppose the Earth had been just slightly different from the way conditions have turned out. It is interesting to speculate what effect small differences might have had on the development of life. (See, for example, G. Benford, "The Hospitable Planet Earth," *Natural History*, December 1973.)

READING MATERIALS FOR THE STUDENT

Books

Cailleux, A. *Anatomy of the Earth* (McGraw-Hill paperback, 1968).

Calder, N. *The Restless Earth* (Viking Press, 1972). [A lavishly illustrated, eloquently written summary of our present knowledge of the Earth's structure and dynamics by a noted science writer.]

Kopal, Z. *The Solar System* (Oxford Univ. paperback, 1973). [Chapter 3 is a concise summary of our present knowledge about our planet.]

Press, F., and R. Siever (eds.). *Planet Earth* (W. H. Freeman and Co., 1974). [Selected reprints from *Scientific American*.]

Articles

Cahill, L. "The Magnetosphere," *Scientific American*, March 1965.

Eberhart, J. "A Magnet Named Earth," *Science News*, May 24, 1975.

Ingham, M. "The Spectrum of the Airglow," *Scientific American*, January 1972.

Jacchia, L. "The Earth's Upper Atmosphere," *Sky and Telescope*, March, April, and May, 1975.

King-Hele, D. "The Shape of the Earth," *Scientific American*, October 1967.

Richter, C. "Earthquakes," *Natural History*, December 1969.

Siever, R. "The Earth," *Scientific American*, September 1975.

Smylie, D., and L. Marshina. "The Rotation of the Earth," *Scientific American*, December 1971.

Webster, W., et al. "A Radio Picture of the Earth," *Sky and Telescope*, January 1975.

Wyllie, P. "The Earth's Mantle," *Scientific American*, March 1975.

READING MATERIALS FOR THE INSTRUCTOR

Akasofu, S., and L. Lazerotti. "The Earth's Magnetosphere," *Physics Today*, December 1975.

Dungey, J. "Some Mysteries Remaining in the Aurora," *Quarterly Journal of the Royal Astronomical Society*, June 1975.

Hartmann, W. *Moons and Planets* (Wadsworth, 1972). [A fine text with more information on the general properties of planets.]

Kuiper, G., and B. Middlehurst (eds.). *Planets and Satellites* (Univ. of Chicago Press, 1961). [Some of the papers in this volume are now significantly dated.]

Roederer, J. "The Earth's Magnetosphere," *Science*, January 11, 1974.

Schubert, G., and O. Anderson. "The Earth's Thermal Gradient," *Physics Today*, March 1974.

van Valen, L. "The History and Stability of Atmospheric Oxygen," *Science*, February 5, 1971.

AUDIO-VISUAL AIDS

Films

The Challenge of Unanswered Questions (1968, color, 15 min). A discussion of the aurorae and how they are studied. (N.A.S.A. film HQ 148.)

Eratosthenes (1969, color, 4 min). A silent film loop, showing how Eratosthenes measured the size of the Earth. (Produced by

Davidson Films; distributed by Bailey Film Associates.)

ERTS--Earth Resources Technology Satellite (1973, color, 27 min). Deals with remote-sensing studies of the surface of the Earth. (N.A.S.A. film HQ 223.)

Flaming Sky (1960, black and white, 27 min). History of our understanding of aurorae. (Part of the Planet Earth Series, distributed by Contemporary Films, McGraw-Hill Co.)

4 RMS--Earth View (1974, color, 28 min). E. G. Marshall narrates this look at the Skylab program. (N.A.S.A. film HQ 239.)

Magnetic Effects in Space (1974, color, 14 min). Dr. O. Garriott performs experiments on Skylab demonstrating the effects of the Earth's magnetic field. (N.A.S.A. film JSC-628.)

The Magnetosphere (1969, color, 27 min). Tells of the discovery, nature, and current investigation of the magnetic zone around the Earth. (Part of "The Sun and the Earth" series, produced by the National Academy of Sciences; distributed by McGraw-Hill Films.)

Mirror in the Sky (1958, black and white, 21 min). The history of our exploration of the ionosphere, with an emphasis on the electronic tools in use today. (Produced by the Educational Foundation for Visual Aids, England; distributed by British Information Services.)

Orbital Photography (1970, color, 30 min). TV program about the various techniques used to monitor the Earth from space. (Part of the Interface Series distributed by the University of California Extension Media Center, Berkeley.)

Research in the Atmosphere (1969, color, 25 min). An historical survey of upper-atmosphere research. (N.A.S.A. film HQ 180.)

This Land (color, about 40 min). A well-made film that traces the physical and biological evolution of the Earth's surface. Magnificent photography and good animation enhance the effectiveness of the presentation. (Produced and distributed by the Shell Oil Company.)

Van Allen Radiation Belts--Exploring in Space (1963, black and white, 17 min). Uses both live and animated photography to describe the exploration of the Earth's radiation belts. (Encyclopaedia Britannica Films.)

SAMPLE EXAM QUESTIONS (MULTIPLE-CHOICE)

1. Eratosthenes measured the size of the Earth by *comparing* an angular distance (determined by observing the same astronomical bodies from two places) with--
   a. the height of the sun on June 21
   b. the depth of the two wells in Egypt
   c. the exact time of noon on June 21
   d. the linear distance between the same two places
   e. the diameter of the entire Earth
2. I have an irregular, purple object that I examine in my laboratory on Earth. If I now take this object to the Moon, which of its physical characteristics will be guaranteed to change?
   a. weight
   b. mass
   c. diameter
   d. density
   e. color

3. Helium and hydrogen molecules can escape from the Earth's atmosphere because they have--
    a. very high molecular weight
    b. high density
    c. higher escape velocity than other molecules
    d. higher mean velocity than other molecules
    e. permission from N.A.S.A.
4. The surface temperature of the Earth is actually warmer than we would expect from the amount of sunlight the Earth receives. This is a result of--
    a. vertical mixing
    b. the force of gravity
    c. the greenhouse effect
    d. the high escape velocity of air
    e. the Earth's magnetic field
5. The process building up mountains on the surface is called--
    a. erosion
    b. plate tectonic activity
    c. seismology
    d. ionospheric bounce
    e. the greenhouse effect
6. Which property of the layers of the Earth's interior can seismologists studying earthquake waves determine?
    a. temperature
    b. magnetic field
    c. density
    d. state (liquid or solid)
    e. more than one of the above
7. The first six most common elements in the Universe were probably in gaseous form on the forming Earth and could escape. The seventh, however, combined with oxygen to form a good part of the Earth's rocks. This element was--
    a. carbon
    b. methane
    c. nitrogen
    d. helium
    e. silicon
8. Aurorae (northern and southern lights) are caused by charged particles traveling along the Earth's--
    a. surface
    b. troposphere
    c. magnetic field
    d. mantle
    e. electric advertising signs
9. A sidereal day, as compared to a solar day, is--
    a. shorter
    b. the same length
    c. longer
    d. variable
    e. This has not yet been measured accurately.
10. The Coriolis force that affects wind patterns is a result of--
    a. the Earth's revolution
    b. the change of seasons
    c. the Earth's rotation
    d. the Earth's magnetic field
    e. the 11-year solar-activity cycle

Answers to Multiple-choice Questions

1. d   2. a   3. d   4. c   5. b   6. e   7. e   8. c   9. a   10. c

SAMPLE EXAM QUESTIONS (ESSAY)

1. Briefly list the layers of the Earth's interior, surface, and atmosphere, starting with the inner core. Describe conditions as we know them in each layer.
2. The composition of the Earth's interior differs significantly from that of the Universe at large. Briefly discuss how we believe this difference came to be.
3. Explain why summer is warmer and winter is colder. Is this effect exactly the same for the northern and southern hemispheres? Why?
4. Astronomers have found that the two most common elements in the Universe are hydrogen and helium. Why is there so little of these gases in the Earth's atmosphere?
5. Explain why you would be incredulous about reports of auroral phenomena in Ecuador.

CHAPTER 4
TIME, PLACE, AND THE SKY

NOTE BY AUTHOR OF TEXT

This is a parenthetical chapter than can be omitted without loss of continuity; some instructors will wish to include this material, others might not. I have tried to minimize the technical complications while making as many everyday connections as could reasonably be included. Instructors who emphasize cultural connections should note the final section, on the calendar.

--I.R.K.

TEACHING NOTES

Instructors who skip this chapter should probably still cover the section on the naming of stars (p. 58), so that the nomenclature in later chapters will make sense to the students.

For courses that include observing sessions, the section on celestial coordinates could be assigned at the proper time.

Much of the material on the sky is presented most effectively in a planetarium. Many planetaria across the country will gladly schedule special programs for visiting classes.

SUMMARY OF KEY POINTS

Coordinate Systems

1. Celestial sphere: an imaginary sphere on which all objects in the sky are taken to lie for convenience.
2. Equatorial coordinates: declination and right ascension are fixed on the celestial sphere; however, the hour angle of an object changes with time.
3. Horizon-system coordinates: altitude and azimuth define positions at a given moment.
4. Sidereal time is defined as the right ascension of the hour circle that lies on the celestial meridian.

Constellations and Star Maps

1. Constellations are 88 precisely defined regions of the sky.
2. Star naming: modern convention is to use the constellation name preceded by a Greek letter.
3. Star positions are usually given using right ascension and declination (with the epoch specified, because precession changes the positions).

Timekeeping

1. The mean solar day has been adopted to average out variations in the length of the day.
2. Today we are on the Gregorian calendar, in which the judicious use of leap years gives us an average number of days in a year very close to the actual period of the Earth's revolution.

Basic Terms Introduced in This Chapter

Celestial sphere
Right ascension and declination
Altitude and azimuth
Sidereal time

TOPICS FOR DISCUSSION AND CLASS ACTIVITIES

1. Discussions of time such as the one in this chapter occasionally generate questions about the nature of time. An excellent film, *Time Is* (1964, color, 30 min; distributed by Contemporary Films, McGraw-Hill Co.), which was produced by S. Toulmin and J. Goodfield, two noted historians of science, addresses itself to this question. See also their book, *Discovery of Time* (Harper & Row paperback). For a more sophisticated look at modern concepts, see *The Physics of Time Asymmetry* by P. Davies (Surrey Univ. Press, 1974).

2. The naming of stars and constellations has a long, involved, and (depending on your tastes) fascinating or tedious history. The classic reference in this field is a volume by R. Allen entitled *Star Names: Their Lore and Meaning*, first published in 1899 and reprinted as a Dover paperback. Many histories of astronomy and astrology dwell on the questions of naming to some extent.

3. The mythological stories associated with the constellations are often used as ways of interesting children in astronomy and the stars. It is the present author's opinion that this is of relatively little value and that perhaps it is time to let these stories and the attendant confusion about the real meaning of constellations die out mercifully. Still, for those interested, an enjoyable little compilation of these stories is *Star Myths and Stories* by P. Proctor (Exposition Press, 1972).

4. The chapter includes a discussion of the "midnight sun" phenomenon at the North Pole, when the sun does not set at all. Students often enjoy speculating about what life would be like in such an environment. This can be a nice prelude to pointing out that on other planets, both in our Solar System and in others, some very fundamental things we take for granted on Earth may be quite different. In this regard, students might like to read the justly famous science-fiction story called "Nightfall" by Isaac Asimov, about a planet in a multiple star system where night comes only once every 2,000 years. The story can be found in the following collections:

   Asimov, I. *Nightfall and Other Stories* (Fawcett/World paperback, 1969).
   Silverberg, R. (ed.). *The Science Fiction Hall of Fame*, vol. 1

(Avon paperback, 1970).

It is worth noting that this story, written in 1941, was recently voted the all-time science-fiction favorite by the science-fiction writers themselves.

5. The measurement of latitude and longitude from astronomical observations, which is briefly described in the chapter, was crucial to the expansion of navigation and trade that ushered in our modern era centuries ago. This is an interesting problem that illustrates the interaction of science with technology and commerce.

REFERENCES AND AUDIO-VISUAL AIDS

Almost any textbook or history of astronomy has a section or two on the topics covered in this chapter. Few authors find any way to make this somewhat dry, rather basic material much more interesting.

Because these are hardly topics that the editors of *Scientific American* would find of breath-taking currency, there are no articles to recommend. Nor are there any films I know of that are worth showing (other than *Time Is*, described in Topic 1 above, which is at best peripherally relevant).

Norton Scientific Co. markets a set of 18 slides on these topics called "Coordinates, Time, and Place." But, all in all, the best visual aid for this chapter is a good planetarium show.

SAMPLE EXAM QUESTIONS (MULTIPLE-CHOICE)

1. The system of coordinates fixed among the stars is called--
   a. alt-azimuth
   b. diurnal
   c. equatorial
   d. meridional
   e. Cartesian
2. A diurnal circle is a circle of constant--
   a. right ascension
   b. declination
   c. altitude
   d. azimuth
   e. daylight
3. The coordinate that measures angular distance around the horizon is called--
   a. declination
   b. right ascension
   c. altitude
   d. azimuth
   e. none of the above
4. The altitude of the pole star is equal to the observer's--
   a. longitude
   b. latitude
   c. azimuth
   d. declination
   e. altitude

5. Which of the following statements is *not* true about the celestial meridian?
   a. It unites the equatorial and horizon coordinate systems.
   b. It divides every diurnal circle in half.
   c. It is fixed among the stars and therefore appears to rotate as the Earth does.
   d. It crosses the zenith.
   e. It touches the north and south points of the horizon.
6. The hour angle of a star is the difference between the location of its hour circle and the location of--
   a. the meridian
   b. the celestial equator
   c. the pole
   d. its diurnal circle
   e. the observer
7. The system of coordinates usually used by modern astronomers to give the position of stars is the--
   a. meridional
   b. horizon
   c. equatorial
   d. Greek
   e. sidereal
8. On the Gregorian calendar, which of the following years is a leap year?
   a. 1700
   b. 1800
   c. 1900
   d. 2000
   e. 1979

Answers to Multiple-choice Questions

1. c 2. b 3. d 4. b 5. c 6. a 7. c 8. d

SAMPLE EXAM QUESTIONS (ESSAY)

1. Describe how the equatorial and horizon coordinate systems are defined. What are the major differences between the two systems? How can the two systems be connected?
2. Define (a) sidereal time, (b) celestial meridian, (c) hour angle, (d) diurnal circle, (e) hour circle.
3. Why is there a need to define a "mean solar day"?
4. What was the reason that the adoption of the Gregorian calendar became necessary? How well did it solve the problems encountered with the Julian calendar?
5. A star of some interest is said by astronomers to be in the constellation of Cygnus the Swan; yet it does not form any part of the swan picture in the sky. How is this possible?
6. After a week of sun, sailing, and beer consumption, you and a friend find yourselves lost at sea. As you begin to sober up, the stars come out. How could you determine your latitude that night, and what equipment would you need to determine your longitude the following day? (Time permitting, you may want to compose a letter of thanks to your astronomy instructor.)

CHAPTER 5
OUR SATELLITE, THE MOON

NOTE BY AUTHOR OF TEXT

Although a chapter about the Moon must necessarily include a lot of descriptive material, I have tried here to keep this to a minimum, emphasizing instead general characteristics and particularly the attempt to understand how and why the Moon got that way. Connections with everyday life are still there in the sections on phases, tides, and eclipses. Note also that this chapter contains the book's only discussion of the space program per se.

—I.R.K.

TEACHING NOTES

While some students find the study of the Moon very interesting, this is another of those chapters that could be omitted without loss of continuity in short courses or courses emphasizing stellar astronomy.

On the other hand, some instructors may want to take advantage of the interest in the Moon generated by the space program. For them, a somewhat longer list of topics, readings, and films is provided.

SUMMARY OF KEY POINTS

Orbital Characteristics

1. Elliptical orbit, with average distance from Earth of about 384,000 km.
2. Because the Earth moves around the Sun, the sidereal month does not equal the synodic month.
3. Rotation is synchronous with revolution.
4. Eclipses occur when the Earth and the Sun are oriented along the Moon's line of nodes.

Physical Characteristics

1. Radius: 1,738 km.
2. Mass: 1/82 of the Earth's mass.
3. Surface gravity: 1/6 of the Earth's.
4. Escape velocity: 2.4 km/sec--too low to keep an atmosphere.
5. Surface features:
   a. maria: basaltic rocks, 3 to 4 billion years old;

b. highlands: breccia, a little over 4 billion years old;
c. craters: smaller impact craters everywhere, but more on the highlands.

6. Interior layers:
   a. crust: about 60 km;
   b. mantle: solid rock;
   c. core: at least partially molten.
7. History: began hot, with rapid surface solidification; later flooding of maria from molten interior; impacts frequent early, less frequent as time goes on.

Tides

1. Moon's pull varies with distance and therefore causes a differential force (which can act more effectively on nonrigid water).
2. Height of the tides varies with the relationship between the Sun and the Moon and with the changing Earth-Moon distance.
3. Tidal friction is slowing down the Earth's rotation and enlarging the Moon's orbit.

Basic Terms Introduced in This Chapter

Perigee and apogee
Librations
Tidal force
Tidal friction
Line of nodes

TOPICS FOR DISCUSSION AND CLASS ACTIVITIES

1. Although the manned exploration of the Moon during the Apollo program added considerably to our understanding of the Moon and the Solar System, many scientists (as well as laypersons) have criticized the expenditure of the huge sums required for the life-support systems. These critics point out that we could have learned almost as much from unmanned space probes (the avenue the Russians have pursued).

   Clearly, the reasons for going to the Moon were not only scientific, but political as well. Was the expense of the Apollo program ultimately justified? This question can spark lively debate, provide a vehicle for a student panel discussion, or lead to some heartfelt student papers.

   N.A.S.A. (see Appendix 2) can of course provide reams of material pointing out the very real benefits of the manned space program for astronomy, medicine, technology, etc. Some other references concerning this controversy are:

   Gold, T. "Machines, Not Men, in Space," *New York Times Magazine*, August 22, 1971.
   Logsdon, J. *The Decision to Go to the Moon* (MIT Press, 1972).
   Mandelbaum, L. "How the U.S. Decided to Go to the Moon," *Science*, February 14, 1969.

   To see what priorities astronomers themselves recommend concerning the exploration of the Universe, see *Astronomy and Astrophysics for the 1970's* [Report of the Astronomy Survey Committee] (National Academy of Science, 1972).

2. Whatever one may think of the cost-effectiveness of the Apollo program, it *has* significantly increased our knowledge of the Moon. Students may be interested in doing a summary report on the Apollo findings. These have been discussed each year at Lunar Science Conferences sponsored by N.A.S.A. and summarized in many articles and several books (see references for this chapter, below).

3. The Moon, because of its proximity, does have significant effects on life on this planet, of which the tides are only one example. Biologists have long been familiar with "lunar rhythms" among certain organisms, the best known of which is the menstrual period of adult women, whose average cycle length is extremely close to the average duration of the lunar cycle. Many organisms--including fireworms, grunions, and salamanders—exhibit behavior timed to lunar phases. (This would be a good report topic for students interested in biology.)

    In addition, psychiatrists and psychologists have recently confirmed folktales of correlations between the phases of the Moon and maxima and minima in tabulations of violent behavior. (The origin of the word "lunatic" is to be found in this correlation.)

4. Although astronomy instructors often take for granted that everyone must have looked at the Moon through a pair of binoculars or small telescope at some time, it is surprising how few students actually have. Therefore, if the opportunity for an evening observing session should arise, don't neglect the Moon as an object of interest. (If you do not have access to a small telescope, local amateur astronomy groups will often be glad to put on a "star party" for your classes.)

5. Even without a telescope, students should be encouraged to keep track of the changing appearance of the Moon as the term goes on. Most compilations of laboratory or observing exercises (see Appendix 3) will have lunar observing projects.

6. It is, of course, an interesting coincidence that the apparent size of the Moon and Sun as seen from our planet are such that *exactly* total eclipses can occur. Recently, Michael Mendillo and Richard Hart of Boston University have calculated whether such a close match occurs for Sun-satellite combinations on any other planet in the Solar System (they did this for 32 satellites, mind you), and no others can have total solar eclipses. From the fact that the only planet whose Moon can produce this impressive phenomenon is also inhabited, the authors infer (with tongue in cheek) that there must be a God. (See *Physics Today*, February 1974, p. 73.) The August 1974 issue of *Physics Today* includes an interesting letter from E. Wicher, pointing out that tidal action is changing the Earth-Moon distance, so that God therefore did not exist until 280 million B.C.

7. Predicting eclipses seems to have been one of the functions of the megalithic structures at Stonehenge. How the stones do this, and what else they can do, makes for an interesting student paper or discussion. See, for example:

Hawkins, G. "Sun, Moon, Men and Stones," *American Scientist*, December 1965.

Hawkins, G. *Stonehenge Decoded* (Delta paperback, 1965).
Hawkins, G. *Beyond Stonehenge* (Harper & Row, 1973).
Hoyle, F. *From Stonehenge to Modern Cosmology* (W. H. Freeman and Co., 1972).

8. Viewing an eclipse firsthand can be an awe-inspiring and thought-provoking experience. Although pictures and verbal descriptions are hardly an adequate substitute, perhaps they can be used to whet the students' appetites to go see the next eclipse in person. Eclipse articles are very abundant in the popular literature of astronomy. Here are a few recent samples:

   Menzel, D., and J. Passachoff. "Solar Eclipse: Nature's Super-spectacular," *National Geographic*, August 1970.
   Silverman, S., and G. Mullen. "Eclipses: A Literature of Mis-adventures," *Natural History*, June/July 1972.

9. The Moon has been a source of great inspiration for poets old and new. There was strong (if not always inspired) reaction from poets to the first Moon landing. Students may want to look at national magazines around the time of Apollo 11 (July 20, 1969) and compare the reactions of scientists, poets, and others. Two good sources for poetic reactions are:

   Phillips, R., (ed.). *Moonstruck: An Anthology of Lunar Poetry* (Vanguard Press, 1974).
   Vas Dias, R., (ed.). *Inside Outer Space* (Doubleday Anchor paperback, 1970).

10. As many students will have noticed, the Moon often looks much bigger near the horizon than it does high in the sky. This "Moon illusion" has been of some interest to psychologists as well as amateur astronomers. For a nice summary, see the article by F. Saunders in *Mercury*, March/April 1976. A more technical reference can be found in an article by F. Restle in *Science*, February 20, 1970.

READING MATERIALS FOR THE STUDENT

Books

Henderson, A., and J. Grey (eds.). *Exploration of the Solar System* (N.A.S.A. publication EP-122, 1974). [$2.05 from U.S. Government Printing Office, Superintendent of Documents, Washington, D. C. 20402.]
Kopal, Z. *The Solar System* (Oxford Univ. Press paperback, 1973). [Chapter 7 is devoted to the Moon.]
Lewis, R. *The Voyages of Apollo* (Quadrangle, 1975).
Mason, B., and W. Melson. *The Lunar Rocks* (Wiley, 1970).

Articles

Allen, J. "Apollo 15: Scientific Journey to Hadley-Appenine," *American Scientist*, March/April 1972.
Driscoll, E. "A Look Inside the Moon," *Science News*, April 7, 1973.
Dyal, P., and C. Parker. "The Magnetism of the Moon," *Scientific American*, August 1971.

Edlington, G., et al. "The Carbon Chemistry of the Moon," *Scientific American,* October 1972.
Faller, J., and J. Wampler. "The Lunar Laser Reflector," *Scientific American,* March 1970.
Goldreich, P. "Tides and the Earth-Moon System," *Scientific American,* April 1972.
Gornitz, V., and R. Jastrow. "Solving the Moon's Riddles," *Natural History,* January 1974. [A nice elementary review of the Apollo findings.]
Levin, E., et al. "The Lunar Orbiter Missions to the Moon," *Scientific American,* May 1968.
Mason, B. "The Lunar Rocks," *Scientific American,* October 1971.
Muehlberger, W., and E. Wolfe. "The Challenge of Apollo 17," *American Scientist,* Nov./Dec. 1973.
Page, T. "Cameras on the Moon with Apollo 15 and 16," *Mercury,* March/April 1972.
Page, T. "The Third Lunar Science Conference," *Sky and Telescope,* March-April 1972.
Page, T. "Note on the Fourth Lunar Conference," *Sky and Telescope,* June, July, and August 1973.
Page, T. "Notes on Lunar Research," *Sky and Telescope,* August 1974.
Science and Public Affairs. Special lunar science issues. November and December 1973. [Contrast these with the pre-Apollo special issues of September 1969. This magazine is also known as the *Bulletin of Atomic Scientists.*]
Watts, R. "Findings from a Sample of Lunar Material," *Sky and Telescope,* March 1970.
Weaver, K. "Have We Solved the Mysteries of the Moon?" *National Geographic,* September 1973.
Wood, J. "The Lunar Soil," *Scientific American,* August 1970.
Wood, J. "The Lunar Rocks," *Smithsonian,* September 1972.
Wood, J. "The Moon," *Scientific American,* September 1975.
Wood, J. "The Moon," in *Man and Cosmos,* ed. J. Cornell and E. Hayes (Norton, 1975).

READING MATERIAL FOR THE INSTRUCTOR

Anderson, D. "The Interior of the Moon," *Physics Today,* March 1974.
Bender, P., et al. "The Lunar Laser Ranging Experiment," *Science,* October 19, 1973.
El-Baz, F. "Surface Geology of the Moon," in *Annual Reviews of Astronomy and Astrophysics,* vol. 12 (1974).
El-Baz, F. "The Moon After Apollo," *Icarus,* August 1975.
Hammond, A. "Lunar Science: Analyzing the Apollo Legacy," *Science,* March 30, 1973.
Hartman, W. *Moons and Planets* (Wadsworth, 1972).
Jeffreys, H. "Tidal Friction," *Quarterly Journal of the Royal Astronomical Society,* June 1975.
Link, F. "Lunar Eclipses," in *Advances in Astronomy and Astrophysics,* vol. 9 (Academic Press, 1972).
Shorthill, R., and J. Saari. "Infra-red Observation on the Eclipsed Moon," in *Advances in Astronomy and Astrophysics,* vol. 9 (Academic Press, 1972).
Taylor, S. *Lunar Science: A Post-Apollo View* (Pergamon Press, 1975).
Wetherill, G. "Of Time and the Moon," *Science,* July 30, 1971.

Volume 8 of *Advances in Astronomy and Astrophysics* (Academic Press, 1971) is devoted entirely to lunar studies.

Special reports on the scientific results from the various Apollo lunar landings appeared in the following issues of *Science:*

Apollo 11: September 19, 1969, and January 30, 1970;
Apollo 12: March 6, 1970;
Apollo 14: August 20, 1971;
Apollo 15: January 28, 1972;
Apollo 16: January 5, 1973;
Apollo 17: November 16, 1973.

AUDIO-VISUAL AIDS

Films

Eclipse of the Quiet Sun (color, 27 min). (N.A.S.A. film NAV-010.)

Eclipses of the Sun and Moon (1966, color, 11 min). Brief introduction to eclipses, using both actual film and animation. (Encyclopaedia Britannica film.)

Man Looks at the Moon (1971, color, 15 min). Historical survey of lunar exploration through the Apollo findings. (Encyclopaedia Britannica Film.)

The Mystery of Stonehenge (1965, color, 57 min). An excellent, dramatic CBS documentary on Stonehenge and its astronomical orientation. The film crew actually tests some of the astronomical correlations proposed by Hawkins and others. (Distributed by Contemporary Films, McGraw-Hill Co.)

Solar Eclipse (1965, color, 15 min). (N.A.S.A. film NAV-009.)

Walk on the Moon (1969, color, 12 min). Brief tracing of the history of rocket flight, beginning with the work of Robert Goddard and culminating on the Apollo Moon landing. [One of the least offensive treatments of the Apollo 11 mission.] (Part of the *Screen News Digest* series, distributed by utility companies and others.)

The list of N.A.S.A. films that deal with lunar exploration is very long—and that is how many of the films seem as well. Most college students today will not hesitate to laugh during the more pretentious moments in these films. However, the films *are* available *free* from your local N.A.S.A. educational center (see Appendix 2). The following is a partial list of current N.A.S.A. films on the moon or the Apollo program:

The Apollo 4 Mission (1968, color, 16 min). (N.A.S.A. film HQ-181.)

The Apollo 5 Mission (1968, color, 17 min). (JSC-461.)

Apollo 10: Green Light for a Lunar Landing (1969, color, 28 min). (HQ-190.)

Apollo 14—Mission to Fra Mauro (1971, color, 27 min). (HQ-211.)

Apollo 15—In the Mountains of the Moon (1971, color, 28 min). (HQ-217.)

Apollo 16—Nothing So Hidden (1972, color, 28 min). (HQ-222.)

Apollo 17—On the Shoulders of Giants (1973, color, 28 min). (HQ-227.)

Assignment: Shoot the Moon (1967, color, 28 min). (HQ-167.)

Debrief: Apollo 8 (1969, color, 28 min). (HQ-188.)

The Eagle Has Landed: Apollo 11 (1969, color, 28 min). (HQ-194.)

The Flight of Apollo 7 (1968, color, 15 min). (HQ-187.)

Hey, Houston . . . We've Got a Problem (1970, color, 28 min). (HQ-200.)

The Lunar Samples of Apollo 11 and 12 (1970, color, 9 min). (JSC-554.)

The Moon, An Emerging Planet (1973, color, 13 min). (HQ-230.)
The Moon, Old and New (1970, color, 25 min). (HQ-209.)
Pinpoint for Science: Apollo 12 (1969, color, 28 min). (HQ-197.)
The Space Dust of Gumdrop and Spider: Apollo 9 (1969, color, 28 min). (HQ-189.)
The Time of Apollo (1974, color, 28 min). (HQ-229.)

Film Loops

A Device to Measure the Moon (1969, color, 4 min). Measuring the Moon's diameter with everyday materials.
Eclipse of the Moon (1964, color, 4 min).
Eclipse of the Sun (1964, color, 4 min).

All three of these loops are available from Bailey Film Associates, 11559 Santa Monica Boulevard, Los Angeles, Calif. 90025.

Slides

Hale, Lick, and Yerkes observatories and various slide-supply firms carry slides of the lunar surface, as seen both through telescopes and by cameras carried by astronauts and space vehicles.

At some N.A.S.A. Educational Services offices, teachers may bring their own film and make copies of slides from the N.A.S.A. collection.

Audio Tapes

Man on the Moon (30 min). R. Lapp and others discuss the political, scientific, and economic implications of the lunar landings. (Distributed by the Center for Cassette Studies.)

SAMPLE EXAM QUESTIONS (MULTIPLE-CHOICE)

1. During which phase of the Moon is it relatively conspicuous during the daytime?
   a. new
   b. full
   c. crescent
   d. quarter
   e. The Moon is *never* visible during the daytime.
2. The point in the Moon's orbit where it is closest to the Earth is called its—
   a. geosync point
   b. perigee
   c. libration
   d. waxing point
   e. apogee
3. Compared to a sidereal month, a synodic month is—
   a. longer
   b. the same length
   c. shorter
   d. variable (sometimes shorter, sometimes longer)
4. When the Moon is between the Sun and the Earth, we call its phase—
   a. full Moon
   b. gibbous
   c. last quarter
   d. new Moon
   e. You can't fool me; the Moon can never get between the Sun and

the Earth.

5. Compared to the period of its revolution about the Earth, the period of the Moon's rotation is—
   a. longer
   b. the same length
   c. shorter
   d. variable (sometimes shorter, sometimes longer)
6. Which of the following characteristics of the Moon is *not* significant in explaining its lack of atmosphere when compared to the Earth?
   a. its mass
   b. its surface gravity
   c. its magnetic field
   d. its escape velocity
7. The Earth is believed to have received a major bombardment at the same time that the Moon did. Yet the Earth does not show extensive cratering. Which of the following is *not* a reason for the absence of craters on Earth?
   a. the Earth's atmosphere
   b. the Earth's Van Allen belts
   c. erosion
   d. plate rearrangements
8. Suppose that, in all the excitement of driving his Moon buggy about, an Apollo astronaut forgot to label one of the rock samples that he brought back to Earth. It is later dated as 4.2 billion years old. Which lunar region is it most likely to have come from?
   a. smooth region in the maria
   b. one of the craters in the maria
   c. the highlands
   d. one of the maria on the backside of the Moon
9. The reason that successive high tides are not exactly one-half day apart is—
   a. the rotation of the Earth
   b. the motion of the Earth around the Sun
   c. the motion of the Moon around the Earth
   d. the Moon's magnetism
   e. the difference between the sidereal and synodic months
10. According to modern theories, what was responsible for the Moon's rotation becoming synchronous with its revolution?
   a. tidal friction due to the Earth
   b. the Earth's Van Allen belts
   c. magnetic forces pulling on the Moon's iron core
   d. the accumulated effect of 4 billion years of eclipses
   e. the initial bombardment that formed the lunar maria

Answers to Multiple-choice Questions

1. d 2. b 3. a 4. d 5. b 6. c 7. b 8. c 9. c 10. a

SAMPLE EXAM QUESTIONS (ESSAY)

1. Why is there a difference between the lengths of the sidereal month and the synodic month? Which do we use for our calendar? Why?
2. Name the various phases of the Moon and describe how the Earth,

Sun, and Moon are aligned during each phase.

3. Briefly describe how the space program has contributed to our understanding of the Moon's structure and history.
4. Outline briefly the history of the Moon's interior and surface, as we understand it today.
5. Describe the various factors that determine the height of tides.
6. Briefly explain when, where, and how each of the following can occur: (a) total eclipse of the Sun; (b) annular eclipse of the Sun; (c) total eclipse of the Moon.

CHAPTER 6
ORBITS AND MOTION

NOTE BY AUTHOR OF TEXT

This is a difficult chapter, both in the concepts that are presented to the student and in the decisions that the teacher must make about how far to go. The basic trouble we face is that everyone grows up with intuitive ideas about motion—ideas that are wrong but are so natural that they were able to delude as brilliant a thinker as Aristotle. I have always regarded the laws of motion as one of the high points of a course in astronomy, where the astronomical problem serves as an entry to some fundamental physics. The level of abstraction is high, however, and the student needs all the help that he can get. I have broken down the laws of motion into simple stages, but they still add up to a more severe exercise than the student will encounter in most other chapters.

—I.R.K.

TEACHING NOTES

This chapter is a miniature physics course in many ways, emphasizing the fundamental laws of motion. If this material is covered at the same pace as the other chapters, some students without physics background will be left confused.

Some instructors may wish to omit parts of this chapter. However, certain sections (such as the determination of masses from Kepler's third law) are quite important to the understanding of many current astronomical techniques and results.

It may be useful to show a film (see "Audo-visual Aids," below) during the study of these concepts. Particularly in the films that feature animation, a good set of drawings may be worth a lot of lecture time.

SUMMARY OF KEY POINTS

1. Kepler's laws for the Solar System:
    a. orbits are ellipses with Sun at one focus;
    b. orbits sweep out equal areas in equal times (or, angular velocity varies inversely with the square of the distance);
    c. the square of the period of the orbit is proportional to the

cube of the average distance from the Sun.

2. Newton's laws of motion:
   a. a body moves with constant velocity unless it is acted on by a net force;
   b. acceleration is proportional to force and inversely proportional to mass;
   c. when two bodies exert forces on each other, the two forces have the same strength but are opposite in direction.
3. Planets moving in circular (or elliptical) orbits require a force to keep their paths curved. (This force is provided by the Sun's gravitation.)
4. Kepler's third law enables us to measure the masses of orbiting bodies in the Universe, which is a crucial determination in many astronomical problems.
5. On the Earth's surface, the acceleration of gravity is the same for all bodies (ignoring air resistance).
6. The orbits of spacecraft are good illustrations of the laws of motion.

Basic Terms Introduced in This Chapter

Kepler's laws
Eccentricity
Perihelion and aphelion; periastron and apastron
Inertia
Vectors
Acceleration
Newton's laws

TOPICS FOR DISCUSSION AND CLASS ACTIVITIES

1. It might be interesting to continue the train of thought begun in discussing Kepler's laws by discussing the differences between moral laws, legal laws, and scientific laws. In particular, how is each arrived at and how is each confirmed, validated, and/or maintained? An excellent little book on a distinguished physicist's view of scientific laws is *The Character of Physical Law* by Richard Feynman (MIT Press paperback, 1965).

2. Galileo, the first person to turn the telescope to the heavens, is also credited with initiating the experimental methods of science that led to many of the principles discussed in this chapter. His life makes a most interesting topic for investigation.

   Begin with the superb film called *Galileo*, read the famous play of the same name by Bertold Brecht, and then continue with the references given in "Reading Materials," below.

3. The chapter refers briefly to some of the physics problems involved with propulsion mechanisms for space travel. This is an excellent discussion or paper topic, particularly in conjunction with thinking about how easy it is for spacecraft from other solar systems to come here. References:

   Bracewell, R. *The Galactic Club* (W. H. Freeman and Co., 1974). [See Chapter 12 in particular.]
   Cameron, A., (ed.). *Interstellar Communication* (Benjamin, 1963).

[See the chapter by Purcell and von Hoerner.]
Purcell, E., et al. "Space Travel: Problems of Physics and Engineering," in *Project Physics Course Reader*, vol. 5 (Holt, Rinehart and Winston, 1971).

4. If the equipment is available, it can be dramatic to demonstrate that the acceleration of gravity on Earth is the same for objects of different mass, particularly if a proverbial feather can be used in an evacuated chamber. Or simply climb (or send a student) to the top of a table and drop two objects of differing mass but similar air-resistance properties. To paraphrase Galileo, one demonstration is worth a thousand dogmatic words.

5. Some students may enjoy looking at the laws discussed in this chapter in the discoverers' own words. McGraw-Hill publishes a set of inexpensive digest translations of important scientific works under the series title *Masterworks of Science*. Volume 1 features Copernicus' *De Revolutionibus* and Newton's *Principia*.
   Michael Hoskin conducts an imaginary TV interview with Newton (as well as Galileo) in his appealing little book, *The Mind of the Scientist* (Taplinger, 1971).

READING MATERIAL FOR THE STUDENT

General

Most physics texts cover the laws discussed in this chapter. The books by Greider, March, and Hewitt are among my favorites for non-science students.

Cohen, I. *The Birth of a New Physics* (Doubleday Anchor paperback, 1960).
Greider, K. *Invitation to Physics* (Harcourt, Brace, Jovanovich, 1973).
Hewitt, P. *Conceptual Physics* (Little, Brown, 1974).
March, R. *Physics for Poets* (McGraw-Hill, 1970).
Project Physics. *Motion in the Heavens* [Project Physics Course Reader, vol. 2] (Holt, Rinehart and Winston, 1970). [A nice collection of articles on Copernicus, Kepler, Newton, and their work.]
Toulmin, S., and J. Goodfield. *The Fabric of the Heavens: The Development of Astronomy and Dynamics* (Harper & Bros., 1961).
Young, L., (ed.). *Exploring the Universe*, 2nd ed. (Oxford Univ. Press paperback, 1971). [A collection of articles on the nature and methods of science, several of which investigate further the laws of motion and their implications.]

Kepler and His Laws

Gee, B. "400 Years: Johannes Kepler," *Physics Teacher*, vol. 9, p. 510 (1971).
Gingerich, O. "Johannes Kepler and the Rudolphine Tables," *Sky and Telescope*, December 1971.
Wilson, C. "How Did Kepler Discover His First Two Laws?" *Scientific American*, March 1972.

Newton and His Laws

Gillispie, C. "Newton with His Prism and Silent Face," in *Science and Ideas*, ed. A. Arons and A. Bork (Prentice-Hall paperback, 1964).

Palter, P., (ed.). *The "Annus Mirabilis" of Sir Isaac Newton (1666-1966)* (MIT Press, 1970). [See especially the second section of papers on his scientific achievements.]

READING MATERIALS FOR THE INSTRUCTOR

For the historical material, the instructor may well profit from the readings listed in the preceding section.

Dreyer, J. *History of Astronomy from Thales to Kepler* (Dover paperback, 1953).

Gingerich, O. *Quarterly Journal of the Royal Astronomical Society*, vol. 13, p. 346 (1972). [A long article on Kepler's work.]

Mintner, J. *Astronautics for Science Teachers* (Wiley Interscience, 1965).

Pannekoek, A. *A History of Astronomy* (Wiley Interscience, 1961).

Roy, A. *Foundations of Astrodynamics* (Macmillan, 1969). [Emphasis on satellites and space travel.]

Textbooks on Celestial Mechanics

Danby, J. *Fundamentals of Celestial Mechanics* (Macmillan, 1962).

Fitzpatrick, P. *Principles of Celestial Mechanics* (Academic Press, 1970)

McCuskey, S. *Introduction to Celestial Mechanics* (Addison-Wesley, 1963).

AUDIO-VISUAL AIDS

Films

A Brief History of Astronomy (1969, color, 9 min). Short, cute animated story of astronomical discovery from Ptolemy to Newton. (Produced by Manitoba Museum of Man and Nature; distributed by McGraw-Hill.)

Galileo: The Challenge of Reason (1969, color, 23 min). Excellent dramatization of Galileo's life and discoveries, as well as the resistance with which his ideas were met. (Distributed by Learning Corp. of America.)

The Law of Gravitation (1964, black and white, 55 min). Film of a public lecture by Nobel-Prize-winning physicist Richard Feynman at Cornell University on the nature of scientific laws in general and the law of gravity as a specific illustration. (Part of the "Character of Physical Law" series, made by the BBC; distributed by Educational Development Center.)

Laws of Motion (1952, color, 15 min). Simple, well-planned demonstrations illustrating Newton's laws and the principle of superposition. (Encyclopaedia Britannica film.)

Newton: The Mind That Found the Future (1970, color, 21 min). Edmund Halley narrates the story of Newton's life and accomplishments; well produced with nice touches of humor. (Distributed by Learning Corp. of America.)

Planets in Orbit: The Laws of Kepler (1960, black and white, 10 min). Brief, elementary film on planetary motion and Kepler's work. (Encyclopaedia Britannica film.)

Satellite Orbits, Satellite Lifetimes (1960, color, 30 min). A more advanced look at satellite dynamics, for those with mathematical background. (Distributed by the Office of Instructional Research, Rensselear Polytechnic Institute.)

Film Loops

Acceleration Due to Gravity (1967, color, 2 loops of 4 min each). The two loops enable students to measure *g*. (Part of the "Motion in the Heavens" series, produced by the National Film Board of Canada; distributed by Holt, Rinehart and Winston.)

Falling Bodies (1966, black and white, 4 min). Slow-motion trace of two falling bodies. (A BBC School TV loop, produced by Gateway Educational Films; distributed by International Communication Films.)

Kepler's Laws (1967, color, 4 min). A computer-animated filmstrip that allows the student to verify Kepler's second and third laws. (Part of the "Motion in the Heavens" series, produced by the National Film Board of Canada; distributed by Holt, Rinehart and Winston.)

SAMPLE EXAM QUESTIONS (MULTIPLE-CHOICE)

1. According to Kepler's second law, the orbit of each planet in equal times sweeps out equal—
   a. sized ellipses
   b. average distances from the Sun
   c. areas
   d. circumferences
   e. numbers of corrupt politicians
2. The semi-major axis of an ellipse is a distance—
   a. along the longer diameter
   b. along the shorter diameter
   c. from focus to focus
   d. along the circumference
3. The eccentricity of a perfect circle is—
   a. zero
   b. one
   c. infinite
   d. pi
   e. cannot be determined
4. Two stars orbit around each other. Their point of closest approach is called—
   a. perihelion
   b. aphelion
   c. periastron
   d. apastron
   e. apogee
5. A hyperbola is—
   a. a very skinny ellipse
   b. an ellipse with zero eccentricity
   c. a very exaggerated statement
   d. an open curve of infinite length
   e. the shape of Jupiter's orbit
6. Compared to your mass on Earth, your mass out in space among the stars would be—
   a. zero

b. negligibly small
c. huge
d. the same
e. This question can't be answered from the information given.

7. As a Saturn V chemical rocket propels an Apollo capsule toward the Moon, what provides the push on the rocket?
   a. the gravitational force of the Earth
   b. the reaction force from the exhaust gases
   c. inertia
   d. the gravitational force of the Sun
   e. the net force of orbital motion
   f. the "hot air" from all those N.A.S.A. press releases
8. According to Newton's laws, a force *must* be acting whenever—
   a. my position changes
   b. time passes
   c. I move with some speed
   d. the direction of my motion changes
   e. more than one of the above
9. The old story about the drunk who sneezed so hard that he fell off his barstool is an exaggerated illustration of what physical law?
   a. law of universal gravitation
   b. law of inertia (Newton's first law)
   c. Kepler's third law
   d. action-reaction law (Newton's third law)
   e. the rule in physics that drunks do not make good scientists
10. The concept of *inertia* as developed by Galileo, Newton, and others tells us that the fundamental state of all matter is—
    a. rest
    b. uniform motion
    c. accelerated motion
    d. none of the above
    e. more than one of the above
    f. California

Answers to Multiple-choice Questions

1. c 2. a 3. a 4. c 5. d 6. d 7. b 8. d 9. d 10. c

SAMPLE EXAM QUESTIONS (ESSAY)

1. Briefly describe each of Kepler's laws. What property of a planet's motion does each law tell us about?
2. Briefly describe each of Newton's laws of motion.
3. Explain how each of Kepler's laws is modified or expanded by Newton's dynamics.
4. How does a rocket travel through empty space if there is nothing to push against? Explain with reference to Newton's laws.
5. Briefly explain why all bodies on Earth fall with the same acceleration.
6. Define the following terms: (a) periastron; (b) apastron; (c) eccentricity (the mathematical meaning, please); (d) acceleration; (e) vector.

CHAPTER 7
THE MOTIONS OF THE PLANETS

NOTE BY AUTHOR OF TEXT

As indicated both here and in the opening of Chapter 6, I have preferred first to present planetary motions as they are, turning only afterward to the question of how they look from the moving Earth. The present chapter is a brief one, serving partly to satisfy curiosity about natural appearances and partly to introduce some phenomena that will be referred to in the Copernican chapter that follows. The concluding paragraphs of this chapter seemed the most suitable place to comment on astrology.

--I.R.K.

TEACHING NOTES

This brief chapter on planetary aspects and orbital elements can also be skipped without loss of continuity.

Unless the various terms and relationships are explained very slowly and carefully, most nonscience students will find this material very confusing and difficult. Especially these days, when many high-schools are cutting back on the number of mathematics courses they require of their graduates, such geometric relationships are so far from the average student's experience that teaching this material can be a very frustrating endeavor.

Once again, visual aids can be very helpful in clarifying these topics. Those with access to a planetarium may want to cover planetary motions there (perhaps at the same time as the material in Chapter 4).

For instructors who want to discuss astrology (or whose students demand such a discussion), some useful references are given under Topic #9 in "Topics for Discussion," Chapter 1.

SUMMARY OF KEY POINTS

1. All the planets move in the same plane and the same direction. The inner planets move faster.
2. The aspects of the planets are different for inferior and superior planets. Thus their visibility and brilliance in the sky vary considerably, quite apart from differences in their intrinsic properties.
3. Six orbital elements determine a planet's path: the semi-major axis, the eccentricity, the inclination, the angle of the

ascending node, the argument of perihelion, and the time of perihelion passage.

4. There is no scientific basis for a belief in astrology.

Basic Terms Introduced in This Chapter

Aspects
Synodic period
Conjunction
Inferior and superior planets
Opposition
Quadrature
Retrograde motion
Orbital elements

TOPICS FOR DISCUSSION AND CLASS ACTIVITIES

1. Students may enjoy looking further into the motion of Venus. For example, it turns out that, between two successive conjunctions, Venus makes exactly four complete rotations with respect to an Earth observer and five complete rotations with respect to a solar observer. Such regularities are believed to be the result of tidal coupling with the Earth. (See, e.g., Chapter 4 of *The Solar System* by Z. Kopal, Oxford Univ. Press paperback, 1973.)

2. Retrograde motion, it is worth mentioning here, is what presented such incredible problems to the ancients seeking to explain the paths of the planets with an Earth-centered system. As discussed in Chapter 8 of the text, this led to the necessity of introducing the whole system of epicycles, which did not disappear until some time after the "Copernican revolution."

3. There may well be fiery objections from some students about the comments on astrology in the chapter. See the discussion and references under Topic #9 in Chapter 1.

   Sometimes it is a worthwhile exercise to take some of the mystery out of the preparation of an astrological horoscope. Michael Zeilik has prepared a laboratory exercise on this subject; see the *American Journal of Physics*, vol. 42, p. 538 (July 1974).

4. One interesting calculation that any introductory physics class can perform is actually to determine the forces the planets exert on a newborn baby. Since neither electromagnetic or nuclear forces will have an effect, the influence must be gravitational. Carl Sagan has suggested that these forces may be profitably compared to the gravitational pull of the obstetrician!

READING MATERIALS AND AUDIO-VISUAL AIDS

There are few books or films appropriate to this chapter. Most textbooks cover this material, some in far greater detail, others in an inconspicuous appendix.

SAMPLE EXAM QUESTIONS (MULTIPLE-CHOICE)

1. The apparent positions of a planet relative to the Sun are given the general name of--
   a. conjunctions
   b. oppositions
   c. aspects
   d. quadratures
   e. inclinations
2. The occasion when a planet (as seen from the Earth) is in line with the Sun and on the far side of it is called--
   a. inferior conjunction
   b. opposition
   c. quadrature
   d. perihelion
   e. superior conjunction
3. When a superior planet is 90 degrees from the Sun, we say it is in--
   a. conjunction
   b. quadrature
   c. opposition
   d. perpendicularity
   e. aphelion
4. For an inferior planet, when is the point of greatest brilliancy as seen from the Earth?
   a. at inferior conjunction
   b. at greatest elongation
   c. at perihelion
   d. at a point between inferior conjunction and greatest elongation
   e. The brightness of inferior planets doesn't vary.
5. A superior planet is most brilliant as seen from the Earth when it is--
   a. at opposition
   b. at quadrature
   c. at conjunction
   d. at greatest elongation
   e. seen through a lot of smog

Answers to Multiple-choice Questions

1. c   2. e   3. b   4. d   5. a

SAMPLE EXAM QUESTIONS (ESSAY)

1. Briefly describe the differences in the aspects of inferior and superior planets.
2. Why is Mercury relatively difficult to see in the evening or morning sky, while Venus is quite conspicuous?
3. Briefly describe the conditions that prevail when an inferior planet appears brightest as viewed from Earth. Do the same for a superior planet. Explain each answer.
4. Explain what causes retrograde motion. Why is the apparent path a loop rather than a line?
5. Briefly describe the six elements that determine an orbit.

CHAPTER 8
THE COPERNICAN REVOLUTION

NOTE BY AUTHOR OF TEXT

To me, the Copernican revolution is one of the great teaching opportunities of a course in astronomy. The narrative carries its own interest, the historic setting relates to other elements of a student's educational experience, and many of its successive steps illuminate the basic nature of science itself. In keeping with this view, I have tried to emphasize the intellectual aspects of the material over the mere narrative. (Note that this chapter is a logically separate unit. An instructor who wished to could omit it—but I think that would be a big mistake.)

—*I.R.K.*

TEACHING NOTES

This chapter is entirely historical in its subject matter, though the issues it raises are most pertinent in our age of increasing doubt about science and of ignorance about its methods.

While skipping the chapter should not impair a study of modern astronomy, many instructors will want to discuss this material for its philosophical as well as its factual value. Although few nonscience students will long remember many of the facts we teach them in a one-semester introductory course, a positive attitude toward free scientific inquiry can hopefully remain with them for life.

The films listed for this chapter are particularly well-made and effective, and they can help make the period come alive for the students.

SUMMARY OF KEY POINTS

1. The "Copernican revolution" was much more a change of attitude rather than merely the emergence of new facts.
2. Ptolemy codified ancient astronomy and constructed a model of eccentric circles upon circles, to represent the motion of the planets as seen from a stationary Earth.
3. During the Middle Ages, much Greek learning was lost. When the ancient writings were rediscovered, Aristotle's philosophy (which included some terrible physics and astronomy based on metaphysics

rather than on actual observations), as "Christianized" by Thomas Aquinas, became the dominant view. Still the Ptolemaic system, which gave better answers, was used for practical calculations.

4. Copernicus simplified the geometric model of planetary motions by putting the Sun in the center and eliminating Ptolemy's awkward equant. However, he still used a form of epicycles and had no *physical* explanation for the movements. Many of his contemporaries (and perhaps Copernicus himself) saw the heliocentric system merely as a computational device and not necessarily as representing physical reality.
5. Giordano Bruno speculated about a much less limited Universe with many inhabited solar systems; he was burned at the stake.
6. Tycho Brahe spent decades in the first careful modern observations of planetary motion (made without a telescope).
7. Kepler used Tycho's data to discover mathematical regularities in the planetary orbits and, eventually, the three laws of orbits that we now associate with his name.
8. Galileo, using a telescope, made a number of observations (phases of Venus, sunspots, the moons of Jupiter) that contradicted contemporary "establishment" views of the Universe, and supported the Copernican system. Galileo also founded the modern science of dynamics by doing careful experiments on the motion of falling bodies.
9. Newton, eventually, brought all this work together by discovering the simple laws that govern all motion (on Earth and in the sky) and thus connecting physics and astronomy.

Basic Terms Introduced in This Chapter

Epicycle
Deferent
Equant

TOPICS FOR DISCUSSION AND CLASS ACTIVITIES

(See also Topics #1, 2, and 5 for Chapter 6.)

1. Our text selects as a key moment in the history of science Kepler's rejection of his early system because of the discrepancy of 8 arc-minutes between theory and observations. This would be a good time to emphasize (once again) to what degree *experiment* is the cornerstone of modern science.

2. Students with a musical background may enjoy investigating two pieces of music inspired by the Copernican revolution:

   Hindemith, P. *Harmony of the World.* [Opera and later symphony, based on the life of Kepler.]
   Smit, L., and F. Hoyle. *Copernicus: Narrative and Credo.* [Available on Desto LP #7178.]

3. There is a thought-provoking little book by Michael Hoskin called *The Mind of the Scientist* (Taplinger, 1971), in which the author "interviews" several great scientists of the past about their work. Both Galileo and Newton are included.

4. The trial of Galileo is only one of a number of "legal" proceedings where scientific ideas were the central question. Students may want to make comparisons with the Scopes "monkey" trial (revolving around the ideas of evolution) or the Oppenheimer security-clearance hearings of the 1950s.

5. Thomas Kuhn, a modern historian of science, has spent a great deal of time and thought examining the Copernican revolution as an archetype of scientific revolutions in general. Among other conclusions, he has proposed that true scientific revolutions alter not only our knowledge about the facts of the world, but our very conception of the methods and structure of scientific inquiry. Motivated students will enjoy reading his *The Copernican Revolution* and *The Structure of Scientific Revolutions* (see "Reading Materials," below).

    A recent conference in honor of the 500th anniversary of Copernicus' birth focused on other revolutions that were as dramatic as the one Copernicus initiated. The transcript is reprinted as *The Heritage of Copernicus: Theories Pleasing to the Mind,* ed. J. Neyman (MIT Press, 1974).

6. Students whose major field of interest is history might like to investigate a new journal published in Britain and edited by Michael Hoskin, called *The Journal of the History of Astronomy.* Each issue features papers on periods ranging from megalithic civilizations to the work of twentieth-century astronomers. (Distributed by Neale Watson Academic Publications, 156 Fifth Ave., New York, N.Y. 10010.)

READING MATERIALS FOR THE STUDENT

General References

Charon, J. *Cosmology* (McGraw-Hill paperback, 1970).

Cohen, I. *The Birth of a New Physics* (Doubleday Anchor paperback, 1960).

de Vaucouleurs, G. *Discovery of the Universe* (Faber and Faber, 1957).

Dreyer, J. *History of Astronomy from Thales to Kepler* (Dover paperback, 1953).

Koestler, A. *The Sleepwalkers* (Grosset & Dunlap paperback, 1963). [A controversial history of astronomy from the Greeks through Newton.]

Kopal, Z. *Widening Horizons* (Taplinger, 1970).

Kuhn, T. *The Copernican Revolution* (Harvard Univ. Press paperback, 1957).

Kuhn, T. *The Structure of Scientific Revolutions,* 2nd ed. (Univ. of Chicago Press paperback, 1970).

Munitz, M. *Theories of the Universe* (Free Press paperback, 1957). [A collection of excerpts from the writings of philosophers and scientists, from ancient days through the 1950s.]

Project Physics. *Motion in the Heavens* [Project Physics Course Reader, vol. 2] (Holt, Rinehart and Winston, 1970). [A collection of readings.]

Young, L., (ed.). *Exploring the Universe,* 2nd ed. (Oxford Univ. Press paperback, 1971). [A collection of readings.]

Ancient Astronomy

Heuer, K. *City of Stargazers* (Charles Scribner's Sons, 1972). [About Aristarchus, Eratosthenes, Hipparchus, and Ptolemy.]

King, J. *The Background of Astronomy* (Braziller, 1958).

(See also books by Charon, Dreyer, and Munitz under "General References," above.)

Copernicus and Bruno

Bronowski, J. "Copernicus the Humanist," *Smithsonian,* April 1973.

Gingerich, O. "Copernicus and Tycho," *Scientific American,* December 1973.

Lerner, L., and E. Gosselin. "Giordano Bruno," *Scientific American,* April 1973.

Neyman, J. "Nicholas Copernicus: An Intellectual Revolutionary," in *The Heritage of Copernicus,* ed. J. Neyman (MIT Press, 1974).

Raman, V. "A Copernicus Quiz," *Physics Teacher,* December 1973.

Ravetz, J. "The Origins of the Copernican Revolution," *Scientific American,* October 1966.

Rosen, E. "Copernicus' Place in the History of Astronomy," *Sky and Telescope,* February 1973.

Tycho Brahe

Christianson, J. "The Celestial Palace of Tycho Brahe," *Scientific American,* February 1961.

Gingerich, O. "Copernicus and Tycho," *Scientific American,* December 1973.

Richardson, R. *The Star Lovers* (Macmillan, 1967). [See chapter on Tycho Brahe.]

Kepler

Casper, M. *Kepler* (Abelard-Schuman, 1959; also Collier Books paperback).

Koestler, A. *The Watershed* (Doubleday paperback, 1960). [The Kepler section from *The Sleepwalkers;* see "General References," above.]

Kopal, Z. *Widening Horizons* (Taplinger, 1970). [Chapter 3 has a nice section on Kepler's work.]

(See also references for Chapter 6.)

Galileo

Drake, S. *Galileo Studies* (Univ. of Michigan Press, 1970).

Drake, S. "Galileo's Discovery of the Law of Free Fall," *Scientific American,* May 1973.

Drake, S., and J. MacLachlan. "Galileo's Discovery of the Parabolic Trajectory," *Scientific American,* March 1975.

Ronan, C. *Galileo* (G. Putnam's Sons, 1974).

Shapere, D. *Galileo: A Philosophical Study* (Univ. of Chicago Press paperback, 1974).

Newton

Richardson, R. *The Star Lovers* (Macmillan, 1967). [Chapter on Newton.]

(See also references for Chapter 6.)

READING MATERIALS FOR THE INSTRUCTOR

Beer, A., and K. Strand (eds.). *Copernicus Yesterday and Today* [vol. 17 of *Vistas in Astronomy*] (Pergamon Press, 1974).
Burtt, E. A. *The Metaphysical Foundations of Modern Science* (Doubleday Anchor paperback, 1954).
Gingerich, O. *Science*, May 10, 1974, p. 660. [An authoritative review of recent Copernicus scholarship.]
Gingerich, O., (ed.). *The Nature of Scientific Discovery* (Smithsonian Institution Press, 1975). [Conference in honor of the 500th Copernicus anniversary, dealing with the Copernican revolution and its consequences.]
Hartner, W. "Galileo's Contributions to Astronomy," in *Vistas in Astronomy*, vol. 11, ed. A. Beer (Pergamon Press, 1969).
Neugebauer, O. "On the Planetary Theory of Copernicus," in *Vistas in Astronomy*, vol. 10, ed. A. Beer (Pergamon Press, 1968).
Toulmin, S., and J. Goodfield. *The Fabric of the Heavens.* (Harper and Bros., 1961).
Whiteside, D. "Before the 'Principia': The Maturing of Newton's Thoughts on Dynamical Astronomy, 1664-1684," *Journal for the History of Astronomy*, vol. 1, part 1 (1970).

AUDIO-VISUAL AIDS

Films

A Brief History of Astronomy (see Chapter 6).
Galileo: The Triumph of Reason (see Chapter 6).
Majestic Clockwork (1974, color, 52 min). Part of the award-winning "Ascent of Man" series written and narrated by Jacob Bronowski, this film discusses the work of Newton and Einstein. (Distributed by Time-Life Films.)
Newton: The Mind That Found the Future (see Chapter 6).
Starry Messenger (1974, color, 52 min). Also part of the "Ascent of Man" series, this film focuses on the various schemes for explaining planetary motion and the conflict over them that led to the trial of Galileo. (Distributed by Time-Life Films.)
An excellent film version of Bertold Brecht's play *Galileo* was made in 1974 by the American Film Theater group; it may be available for general distribution in a few years.

Film Loops

Retrograde Motion—Geocentric Model (1967, color, 4 min).
Retrograde Motion—Heliocentric Model (1967, color 4 min).
In these two film loops, large mechanical models are used to show the motion of Mars as viewed from a fixed position above the plane of the Solar System and from the Earth. (Distributed by Holt, Rinehart and Winston.)

Slides

A set of slides on the history of astronomy, including diagrams of the various theories and portraits of the scientists involved, is available from Norton Scientific Products, for example.

Audio Tapes

The Evolution in Theories of Gravitation (9 min). The distinguished physicist Robert Dicke discusses the history of our ideas about gravity. (Available from Spring Green Multimedia.)

The Newtonian Universe (60 min). A discussion among F. Priestley, J. McGuire, and G. Buchdahl about Newton's cosmology, its interpretation, and its effects. (Distributed by the Center for Cassette Studies.)

The Renaissance Cosmos (60 min). S. Drake and S. Zitner discuss the Copernican revolution, its times, and its place in our culture. (Distributed by the Center for Cassette Studies.)

SAMPLE EXAM QUESTIONS (MULTIPLE-CHOICE)

1. In Ptolemy's system, the planets themselves moved along secondary circles called—
   a. equants
   b. epicycles
   c. deferents
   d. hyperbolas
   e. stadia
2. In Ptolemy's system, the Earth moved along a circle called—
   a. an equant
   b. a deferent
   c. an epicycle
   d. an ellipse
   e. The Earth did not move in Ptolemy's system.
3. The ancient Greek thinker whose views came to dominate the medieval world was—
   a. Archimedes
   b. Aristotle
   c. Ptolemy
   d. Aquinas
   e. Zorba
4. Which of the following statements about Copernicus' system of planetary motions is correct?
   a. It did away with the need for epicycles of any kind.
   b. It still had the Earth in the center of the Solar System.
   c. It used uniform circular motion.
   d. It was the first time in the history of thought on Earth that anyone had regarded the Sun as the center of the Solar System.
   e. It included a *physical* explanation of why the planets moved.
5. The great observational astronomer who left behind him a 20-year record of the motions of the planets made without a telescope was—
   a. Giordano Bruno
   b. Nicholas Copernicus
   c. Tycho Brahe
   d. Johann Kepler
   e. Galileo Galilei
6. The mathematician who found regularities in the planetary observations made by someone else, and who proposed several laws of planetary orbits, was—
   a. G. Bruno
   b. N. Copernicus
   c. T. Brahe
   d. J. Kepler
   e. I. Newton

7. Which of the following did Galileo *not* discover with his new telescope?
   a. phases of Venus
   b. moons of Jupiter
   c. moons of Mars
   d. sunspots
   e. mountains on the Moon
8. How was Galileo able to slow down motion sufficiently to show that all objects on Earth fall with the same acceleration?
   a. by climbing to the top of a leaning tower
   b. by using an evacuated chamber
   c. by using special apples from Newton's orchard
   d. with a special pulley-and-rope arrangement
   e. by using inclined planes
9. At first, when Newton worked out the orbit of the Moon as predicted by his law, the numbers did not agree with observations. The key measurement that was later made and brought the numbers into agreement was—
   a. the roundness of the Earth
   b. the radius of the Earth
   c. the distance to the Moon
   d. the distance to the Sun
   e. the radius of the Moon

Answers to Multiple-choice Questions

1. b 2. e 3. b 4. c 5. c 6. d 7. c 8. e 9. b

SAMPLE EXAM QUESTIONS (ESSAY)

1. Contrast Ptolemy's system of the planetary motions with that of Copernicus. How were they similar, and where did they differ? In particular, what did each say about (a) the position of the Earth, (b) the kinds of circles needed to represent the motions, (c) the points with respect to which motion was uniform, and (d) the physical cause of the motions?
2. What were the factors that made the Copernican revolution possible? In particular, how was the world of Copernicus different from that of Aquinas?
3. Discuss how and why Tycho Brahe's work was a necessary prelude to the discoveries of Johann Kepler.
4. List four observations Galileo made through his new telescope, and briefly explain how each contradicted commonly held views of his time.
5. Discuss how Newton's work is the culmination of the Copernican revolution. How did Newton's work build upon the work of his predecessors, and in what ways did it differ from their approaches?

CHAPTER 9
THE PLANETS

NOTE BY AUTHOR OF TEXT

This long chapter has much factual detail, but its point of view is always centered on the gathering and interpretation of observations. In the interpretation, a lot of new physics makes its way in; as I explained in my comments on Chapter 2, it seemed better to introduce the physics bit by bit wherever it was needed. Note particularly the final section, which tries to pull everything together.

—I.R.K.

TEACHING NOTES

This chapter is a rapid survey of the Solar System. Some teachers may want to spend a great deal more time describing conditions on and around the various planets, although this can usually be done only by skimping on the later chapters.

The last section, on the chemical selectivity of planet formation at different temperatures, will be somewhat more difficult for non-science students unused to keeping several variables in mind simultaneously.

As the space program proceeds, some of the material in this chapter naturally will be out-dated. Students (and instructors) should refer to current issues of such magazines as *Science News*, *Sky and Telescope*, or *Mercury* for the latest information about the Solar System.

SUMMARY OF KEY POINTS

Terrestrial Planets

1. *Mercury* is small and very hot, has no atmosphere, and is cratered like the Moon. Its rotation rate is two-thirds of its orbital period (as measured by the Doppler shift of radar waves).
2. *Venus* is the same size and density as the Earth, but much hotter because of the carbon-dioxide greenhouse effect, and with a much thicker atmosphere. Its rotation is retrograde and slower than its orbital period.
3. *Mars* is small, with a thin atmosphere composed mainly of carbon dioxide. Mariner 9 sent back the first clear views of the sur-

face, including volcanoes, long canyons, and stream beads. These last may be evidence that in other times Mars may have had flowing water. Mars has two tiny, irregularly shaped moons.

Jovian Planets

1. *Jupiter* is large, but with low density throughout and no sharp surface. There is a powerful magnetic field, generated in the planet's core, which is probably liquid metallic hydrogen. Charged particles trapped by this field generate significant amounts of radio radiation. It has 14 satellites, some of which may be captured asteroids.
2. *Saturn* is similar to Jupiter, although less dense, and has a system of rings. One of its satellites, Titan, has an atmosphere.
3. *Uranus,* identified by chance in 1781, and *Neptune,* discovered in 1846, are both giants similar to, though smaller than, Jupiter. Uranus has an axis of rotation perpendicular to the plane of its orbit.

Pluto

1. Discovered in 1930, *Pluto* has an eccentric orbit that can sometimes bring it closer to the sun than Neptune, and which is synchronized with Neptune's orbit in a stable resonance. Its diameter is probably as small as that of Mars or Mercury, and is therefore difficult to measure.
2. Pluto may once have been a satellite of Neptune.

The Family of Planets

1. Many of the large-scale differences among the planets can be explained as a consequence of the effect of different temperatures on the various bodies as they were forming.
2. In the inner Solar System, the temperature was sufficiently high that many lighter elements and their compounds did not condense out and were lost. Because these elements are thought to have made up a high proportion of the proto-planetary material, the inner planets are significantly smaller.
3. In the outer Solar System, where temperatures were lower, much more material could condense. This, in turn, enabled the larger planets that formed to hold on to even more material, because of the resulting higher gravitational pull.

Basic Terms Introduced in This Chapter

Spectral absorption lines
The absolute scale of temperature
Stable resonance

TOPICS FOR DISCUSSION AND CLASS ACTIVITIES

1. To demonstrate the amazing degree to which our knowledge of the Solar System has increased in recent years, it may be instructive to have students compare the information in the text to descriptions of the planets in books or encyclopedias published 20 or more years ago.

2. Teachers interested in seeing some vivid examples of how the material in this chapter can be "livened up" without sacrificing

scientific accuracy ought to read Part Two of Carl Sagan's *The Cosmic Connection* (Dell paperback, 1973). Among other subjects, Sagan traces how the red-hot, choking, sulfurous environment of Venus, replete with pools of molten metal, reminds him of a place much celebrated in religious literature. He also proposes rules for a very slow, one-man game of Phobos baseball.

3. The controversy in the early 1900s over the so-called "canals of Mars" is an excellent illustration of some of the problems and pitfalls of science. The features on the red planet were both hard to see and easy to misinterpret. Good discussions can be found in the following sources:

   Hetherington, N. "Lowell's Theory of Life on Mars," in *Astronomical Society of the Pacific Leaflet* #501 (March 1971).
   Shklovsky, K., and C. Sagan. *Intelligent Life in the Universe* (Delta paperback, 1966). [See Chapter 20.]
   Struve, O., and V. Zebergs. *Astronomy in the 20th Century* (Macmillan, 1962). [See Chapter 8.]

4. Whereas much of the information in Chapter 5 about the Moon came from the *manned* landing missions, most of the recent discoveries about the planets have been the result of *unmanned* probes. The advantages and disadvantages of each method make for some interesting discussion.

5. It is interesting that the existence of the two moons of Mars was "predicted" by Jonathan Swift in *Gulliver's Travels* about 150 years before Asaph Hall discovered them in 1877 (see Chapter 26 of the book by Shklovsky and Sagan listed in Topic #3, above). Lyle Boyd has taken this as the starting point for a marvelous proof that Swift must have been a Martian (*Graduate Journal*, Winter 1965).

6. For teachers who wish to make the Martian surface more vivid for their students, the National Geographic Society has published a large map of Mars (based on the Mariner 9 data). The map was included in the February 1973 issue of *National Geographic*. For a three-dimensional model of Mars, see the film *Mars: The Search Begins* ("Audio-Visual Aids," below).

7. The history of the observations of Jupiter's red spot makes an interesting topic for a student report. Good articles can be found in *Scientific American* (February 1968) and *Sky and Telescope* (May 1968). Many of the recent works cited in "Reading Materials" (below) include the current theory about the spot, based on the Pioneer fly-by data.

8. The text mentions that, in its polar region, Uranus has a 42-year day and 42-year night. It is instructive to have students think about other unusual phenomena as they learn about the Solar System. How many of the things we take completely for granted would change dramatically if we were living on one of the other planets?

9. The measurement of Pluto's diameter is a good example of an excruciatingly difficult, yet quite fundamental observation.

Ingenious techniques are required, together with patient observations. Throughout the course, students may enjoy delving more deeply into the details of how such measurements are carried out. *Sky and Telescope* magazine, which has a good index in each December issue, is a fine place for them to start. For more on Pluto's diameter, see (for example) the October 1965 issue.

10. As discussed in the text, several significant differences between the Earth and Venus are the result of the difference in temperature caused by the Earth's greater distance from the Sun. Because Venus is clearly uninhabitable for our species, it is interesting to speculate how narrow the habitable zone in our Solar System really is. I. Rasool has suggested that, if the Earth had been only 6% closer to the Sun, there might very likely be no one here to take (or give) astronomy courses.

READING MATERIALS FOR THE STUDENT

General Books on the Solar System

Cornell, J., and E. Hayes (eds.). *Man and Cosmos* (Norton, 1975). [Edited papers from a Smithsonian lecture series on the Solar System.]

Henderson, A., and J. Grey (eds.). *Exploration of the Solar System* (N.A.S.A. Publication EP-122, 1974). [A concise summary of current results from and prospects for space exploration. Available for $2.05 from U.S. Government Printing Office, Superintendent of Documents, Washington, D. C. 20402.]

Kopal, Z. *The Solar System* (Oxford Univ. Press paperback, 1973). [The style is a little dry, but this is an excellent summary of what we know and how we know it.]

Sagan, C. *The Cosmic Connection* (Dell paperback, 1973). [A number of the brief, often speculative essays in this well-written book deal with the current view of our Solar System.]

Sagan, C., and J. Leonard. *Planets* (Life Science Library, 1966). [Although some parts of this book are now dated, the clear writing and lavish illustrations still make this a fine layperson's introduction to the planets.]

Scientific American. The Solar System (W. H. Freeman and Co., 1975). [A reprint of the September 1975 issue of the magazine, with up-to-date articles on each planet.]

General Articles on the Solar System

Alexander, J. "New Vistas in Planetary Radio Astronomy," *Sky and Telescope*, March 1976.

Grossman, L. "Condensation of the Solar System," in *McGraw-Hill Yearbook of Science and Technology* (McGraw-Hill, 1974).

Lewis, J. "The Chemistry of the Solar System," *Scientific American*, March 1974. [Reprinted in *New Frontiers in Astronomy*, ed. O. Gingerich, W. H. Freeman and Co., 1975.]

McGuire, J., et al. "The Size of the Solar System," *Scientific American*, April 1961.

"Radar Probes of the Planets," *Science News*, August 24/31, 1974.

Sagan, C. "The Solar System," *Scientific American*, September 1975.

Sagan, C. "The Climates of Planets," in *Science Year 1975* (World Book Encyclopedia).

Shapiro, I. "Radar Observations of the Planets," *Scientific American,* July 1968.

Mercury

Beatty, J. "Mariner 10's Second Look at Mercury," *Sky and Telescope,* November 1974.

Driscoll, E. "On the Heels of Mercury," *Science News,* October 6, 1973.

Hartmann, W. "The Significance of the Planet Mercury," *Sky and Telescope,* May 1976.

"Mercury: Revising the Textbooks," *Science News,* April 20, 1974.

"Mercury Revisited by Mariner 10," *Sky and Telescope,* May 1975.

Murray, B. "Mercury," *Scientific American,* September 1975.

Strom, R. "The Planet Mercury as Viewed by Mariner 10," *Sky and Telescope,* November 1974.

Weaver, K. "Mariner Unveils Venus and Mercury," *National Geographic,* June 1975.

Venus

Driscoll, E, "Our Cratered Sister Planet," *Science News,* August 4, 1973.

James, J. "The Voyage of Mariner 2," *Scientific American,* July 1963. [Now mainly of historical interest.]

"New Radar Maps of Venus," *Sky and Telescope,* November 1970.

Rasool, S. "Venus, Star of Sweet Confidences," *Natural History*, June/July 1969.

Sagan, C. "The Climates of the Planets," in *Science Year 1975* (World Book Encyclopedia).

"The Turbulent Veils of Venus," *Science News,* February 16, 1974.

"Veneras 9 and 10 on Venus," *Sky and Telescope,* December 1975.

"Venus Observed by Mariner," *Sky and Telescope,* April 1974.

Weaver, K. "Mariner Unveils Venus and Mercury," *National Geographic,* June 1975.

Young, A., and L. Young. "Venus," *Scientific American,* September 1975.

Mars

"The Argon on Mars: Portents, Puzzles," *Science News,* June 21, 1975.

Capen, C., and L. Martin. "Mars' Great Storm 1971," *Sky and Telescope,* May 1972.

Carr, M. "The Volcanoes of Mars," *Scientific American,* January 1976.

Cromie, W. "Mars Is an Active Planet," in *Science Year 1973* (World Book Encyclopedia).

Driscoll, E. "Mariner's Intriguing Evidence of a Formerly Watery Mars," *Science News,* March 10, 1973.

Hartmann, W. "If Mars Once Had Rivers, Was the Sun Then Hotter?" *Smithsonian,* September 1975.

Hartmann, W., and O. Raper. *The New Mars: The Discoveries of Mariner 9* (N.A.S.A. Publication SP-337, 1974). [The authoritative review of the mission results.]

Hartmann, W., et al. "The New Mariner 9 Map of Mars," *Sky and Telescope,* August 1972. [See also the issue of April 1972.]

Moore, P., and C. Cross. *Mars* (Crown, 1973). [A beautiful, but overpriced, annotated atlas of the red planet.]

Murray, B. "The Surface of Mars," *Scientific American,* May 1970.

Murray, B. "Mars from Mariner 9," *Scientific American,* January 1973.

Murray, B., (ed.). *Mars and the Mind of Man* (Harper & Row, 1973). [A marvelous layperson's summary of our picture of Mars after the Mariner 9 mission, with thoughtful discussion of the social and

READING MATERIALS FOR THE INSTRUCTOR

Ash, M., et al. "The System of Planetary Masses," *Science*, December 12, 1971.

Cameron, A. "The Outer Solar System," *Science*, May 18, 1973.

Fox, K. "Planetary Atmospheres: How Do We Know?" *Physics Teacher*, May 1973.

Hartmann, W. *Moons and Planets* (Wadsworth, 1972). [A fine introductory text about the Solar System for the nonspecialist.]

Hunten, D., (ed.). *The Atmosphere of Titan* (N.A.S.A. Publication SP-340, 1974). [Proceedings of a 1973 workshop.]

Journal of Geophysical Research. Special issue devoted to Mercury. June 10, 1975.

Lewis, J. "The Atmosphere, Clouds, and Surface of Venus," *American Scientist*, Sept./Oct. 1971.

Owen, T. "The Atmosphere of Jupiter," *Science*, March 27, 1970.

Smoluchowski, R. "Jupiter 1975," *American Scientist*, Nov./Dec. 1975.

Young, R., and D. DeVincenzi. "From Mars with Love," *Science*, November 8, 1974.

The Mariner 9 results were reported in *Science*, January 21, 1972, and September 14, 1973. The Mariner 10 results were reported in *Science*, July 12, 1974. The Pioneer 10 observations are summarized in *Science*, January 25, 1974; those of Pioneer 11 in *Science*, May 2, 1975.

AUDIO-VISUAL AIDS

Films

Close-up of Mars (1966, color, 30 min). Don Herbert (who played Mr. Wizard) narrates this well-done TV program about the design and flight of the Mariner 4 mission to Mars. (Part of the "Experiment Series," distributed by Indiana University.)

Jupiter Odyssey (1974, color, 28 min). Nicely done, thought somewhat over-produced, story of the Pioneer 10 mission and the first results. (N.A.S.A. film HQ-243.)

Mariner--Mars '69 (1971, color, 21 min). A Mariner 9 summary; not as good as *Mars: The Search Begins*. (N.A.S.A. film HQ-208.)

Mars Minus Myth (1973, color, 18 min). Bruce Murray narrates this fine post-Mariner film. (Distributed by Churchill Films.)

Mars: The Search Begins (1973, color, 27 min). An excellent, well-produced summary of the Mariner 9 results, with narration by Carl Sagan and other project scientists. (N.A.S.A. film HQ-236.)

A Mission for Mariner (1969, color, 14 min). Animated summary of the 1967 Mariner flyby of Venus. (N.A.S.A. film HQ-192.)

Other Planets--No Place Like Earth (1971, color, 30 min). Based on the Time-Life book, *The Planets*, this is a somewhat updated tour of the Solar System, making use of computer-animation techniques. (Distributed by Time-Life Films.)

Red Planet Mars (1968, color, 17 min). A history of Martian "exploration" from 1672 through the 1965 Mariner 4 mission. (Distributed by the International Film Bureau.)

Film Loops

Evolving Views of Mars (late 1960s, color, 3 min). A silent loop showing naked-eye, telescope, and satellite views of the planet,

esthetic implications of space exploration.]
Pollack, J. "Mars," *Scientific American,* September 1975.
Sagan, C. "Of Mars, Martians and Mariner 9," *Horizon,* Summer 1973.
Sagan, C. "The Climates of the Planets," in *Science Year 1975* (World Book Encyclopedia).
Weaver, K. "Journey to Mars," *National Geographic,* February 1973.

As this resource book goes to press late in 1976, the Viking mission results are just coming in. Clearly, these findings will supersede many of the articles listed above. Check recent issues of *Scientific American, Sky and Telescope, Science News,* and *National Geographic* for reports on the Viking findings.

Jupiter

Chapman, C. "The Discovery of Jupiter's Red Spots," *Sky and Telescope,* May 1968.
Eberhart, J. "Jupiter in (Sharper) Focus," *Science News,* May 10, 1975.
Finnel, R., et al. *Pioneer Odyssey: Encounter with a Giant* (N.A.S.A. Publication SP-349, 1974). [A nontechnical review of the Pioneer 10 flyby mission.]
Franklin, K. "Radio Waves from Jupiter," *Scientific American,* July 1964.
Gehrels, T. "Flyby of Jupiter," *Sky and Telescope,* February 1974.
Hubbard, W., and J. Jokipii. "New Studies of Jupiter," *Sky and Telescope,* October 1975.
Kowal, C. "All the Outer Satellites of Jupiter," *Sky and Telescope,* April 1976. [Photographs of VI through XIV.]
Millis, R. "The Galilean Satellites," *Mercury,* Jan./Feb. 1974.
"Pioneer 11: Through the Dragon's Mouth," *Sky and Telescope,* February 1974.
"Refined Close-ups of Jupiter," *Sky and Telescope,* July 1975.
Simmons, T. "Mighty Jupiter Could Be a Star That Didn't Make It," *Smithsonian,* September 1974.
Watts, R. "Pioneer Observes Jupiter," *Sky and Telescope,* February 1974.
Wolfe, J. "Jupiter," *Scientific American,* September 1975.

Saturn and Titan

Benford, G. "Atmospheric Titan," *Natural History,* April 1974.
Cruikshank, D., and D. Morrison. "Titan and Its Atmosphere," *Sky and Telescope,* August 1972.
Goodman, J. "The Edgewise Presentation of Saturn's Rings," *Sky and Telescope,* September 1965.
Hunten, D. "The Outer Planets," *Scientific American,* September 1975.
"Radar Echoes from Saturn," *Sky and Telescope,* April 1973.
Sagan, C. "Titan," *Astronomy,* March 1975.
Veverka, J. "Measuring the Sizes of Saturn's Satellites," *Sky and Telescope,* December 1975.

The Outer Planets

Gingerich, O. "The Solar System Beyond Neptune," *Scientific American,* April 1959.
Hunten, D. "The Outer Planets," *Scientific American,* September 1975.
"Pluto's Mass and the Motion of Neptune," *Sky and Telescope,* February 1969.
Sinton, W. "A Near Infra-red View of the Uranus System," *Sky and Telescope,* November 1972.

through the Mariner 7 results. (Produced and distributed by the American Association of Physics Teachers.)

Jupiter Satellite Orbit (1967, color, 4 min). Shows the motion of the satellites around Jupiter as seen through a 24" telescope. (Distributed by Holt, Rinehart and Winston.)

Mars and Jupiter (1969, color, 4 min). Time-lapse film of Mars and Jupiter through a telescope. (Distributed by Bailey Film Associates.)

Report from Mars: Mariner 9 (early 1970s, color, 3 min). Silent loop of the TV pictures of about 40% of the Martian surface. (Produced and distributed by the American Association of Physics Teachers.)

Slides

Telescope views of the closer planets can be obtained from the Hale, Lick, and Yerkes observatories, as well as other distributors. For example, the Astronomical Society of the Pacific markets a set with color views of Mars, Jupiter, Saturn, and a comet.

As mentioned for Chapter 5, some N.A.S.A. Educational Services Offices (Appendix 2) will let instructors duplicate slides from the space program. At the very least, any office can tell you what commercial firms have been authorized to distribute the photographic results from recent planetary probes.

Audio Tapes

Man, the Planets and the Future (54 min). Dr. Bruce Murray discusses the implications of our exploration of the Solar System. (Distributed by the Center for Cassette Studies.)

The Search for Life on Mars (20 min). An interview with Dr. Harold Klein, chief project scientist for the Viking mission. (Astronomical Society of the Pacific radio-program cassette ER-136.)

Weather on the Planets (30 min). An interview with Dr. Von Eschelman about planetary atmospheres. (Astronomical Society of the Pacific radio-program cassette ER-145.)

SAMPLE EXAM QUESTIONS (MULTIPLE-CHOICE)

1. How were astronomers able to measure the rotation rate of Mercury?
   a. by watching large craters with the best high-mountain telescopes as the planet turned
   b. by watching the Earth's Moon, which has exactly the same rotation rate
   c. by bouncing radar beams from Mercury
   d. The rotation rate did not have to be measured; it could easily be calculated from Kepler's laws.
   e. The rotation rate of Mercury has never been measured; the planet is simply too close to the Sun.
2. The observation that the surface of Venus is significantly hotter than the Earth's can be explained—
   a. completely by the fact that Venus is closer to the Sun
   b. mostly by the huge greenhouse effect of Venus' atmosphere
   c. by the fact that Venus rotates very slowly
   d. by the fact that Venus has no moon to draw off heat
   e. only by writing to the Venus Convention and Visitors' Bureau
3. Indications are that the dark side of Venus is almost as hot as the sunlit side. Astronomers believe that the temperature is

evened out by—

a. winds
b. rocks on the surface
c. oceans of liquid sulfuric acid
d. the planet's fast motion around the Sun
e. none of the above

4. What is the current interpretation of the seasonal color changes long observed on the Martian surface?
   a. volcanic eruptions, spewing out huge amounts of lava
   b. growth and decay of Martian vegetation
   c. freezing of the Martian seas during the cold Martian winter
   d. movement of wind-blown dust
   e. blooming of lichens during the Martian spring
5. The reason that the southern polar cap on Mars disappears completely during the southern summer, while the northern cap only shrinks but never disappears, is that—
   a. Mars has no atmosphere at all
   b. the polar caps are made of carbon dioxide rather than water
   c. the orbit of Mars is very eccentric, and Mars is much closer to the Sun during the southern summers than during the northern ones
   d. streams keep the northern polar cap constantly resupplied with material for ice
   e. the Martian Olympics are always held in the northern hemisphere
6. Of the planets in the Solar System, which one would float in your bathtub (if it would fit)?
   a. Mercury
   b. Venus
   c. Mars
   d. Jupiter
   e. Saturn
7. The interior of Jupiter probably consists of—
   a. iron and nickel
   b. rocks, like the surface
   c. methane and ammonia gas
   d. liquid hydrogen and helium
   e. charged particles in a magnetic field
8. The reason that ammonia is so much less prominent in the spectrum of Saturn than in that of Jupiter is that—
   a. there is almost no ammonia at all on Saturn
   b. Saturn does not have a red spot to stir up ammonia
   c. Saturn is cooler than Jupiter, and the ammonia condenses out
   d. Saturn has rings rather than colored bands
   e. Saturn is much more dense than Jupiter, and the ammonia could not form
   f. there is less house-cleaning needed on Saturn
9. The characteristic that distinguishes Titan from most moons in the Solar System is its—
   a. magnetic field
   b. location
   c. composition
   d. atmosphere
   e. rings
10. The characteristic that makes Uranus a unique planet in the Solar System is its—
    a. low temperature

b. axis of rotation
c. atmospheric composition
d. orbital path around the Sun
e. calculated density

11. Which of the following is a true statement about the atmosphere of the Earth that you are breathing while taking this exam?
    a. It has about as much carbon dioxide in it as the atmosphere of Venus.
    b. It is very similar to the atmosphere of Mercury, which is also a terrestrial-type planet.
    c. It is not the original atmosphere with which the Earth began.
    d. It is similar to the composition of the Earth's core, because they both formed from the same material.
    e. none of the above

Answers to Multiple-choice Questions

1. c 2. b 3. a 4. d 5. c 6. e 7. d 8. c 9. d 10. b 11. c

SAMPLE EXAM QUESTIONS (ESSAY)

1. How did the advent of space astronomy change our understanding of the Solar System? Discuss three specific observations made by satellites that could not have been made from the surface of the Earth.
2. List three specific characteristics that are different in the terrestrial and in the Jovian planets. How can these differences be explained?
3. Explain briefly how we have measured the rotation rates of Mercury, Venus, and Jupiter. How do our techniques and findings differ for each of these planets?
4. An amateur astronomer, after 100 nights of patient observation, announces that he has discovered a tenth planet, between the orbits of Neptune and Pluto. Discuss why you should be skeptical about such a discovery.
5. According to several theories of Solar System formation, all the planets in the Solar System should have begun with about the same size. What happened?
6. If all the planets formed from the same material, why is the atmosphere of the Earth different from that of Jupiter?

CHAPTER 10
THE OTHER BODIES OF THE SOLAR SYSTEM

NOTE BY AUTHOR OF TEXT

As in the preceding chapter, the factual detail is introduced here from an observational and interpretative point of view. In a short course much of this material can be skipped, although some acquaintance with asteroids and comets is desirable as background for the discussion of the origin of the Solar System in Chapter 11.

—I.R.K.

TEACHING NOTES

Rather than skipping the chapter entirely, some instructors may prefer to give these subjects an extremely brief mention. Most students wonder about shooting stars and, after the controversy and publicity that surrounded Comet Kohoutek a few years ago, comets also evoke a great deal of student interest. This material is cheeringly easy to grasp, so that a single session or less on Solar System debris may suffice.

SUMMARY OF KEY POINTS

Asteroids

1. Asteroids orbit the Sun between Mars and Jupiter; they vary in size, from 800 km to tiny chunks. About 1,800 have been catalogued. Asteroid orbits can be perturbed by Jupiter; the Trojan group has become locked into an orbit with the same period as Jupiter.
2. Asteroids have low reflectivity, and therefore probably are made of rock. For all but the largest ones, their gravitational forces are weaker than the tensile strengths of the rocks; therefore their shapes are irregular.

Comets

1. The comets we observe are small members of the Solar System with orbits so large that they spend most of their time very far away from the inner Solar System.
2. When a comet approaches the Sun, the ices (and, eventually, the other materials) of which it is composed vaporize, producing the observable coma and tail. The tail is "blown" away from the comet

by the Sun's radiation pressure and by the solar wind.

3. Because each passage near the Sun depletes a comet, the existence of comets today implies that there is probably a large reservoir of comets beyond Pluto.
4. Periodic comets (those with short periods) have had their orbits changed by the planets. Halley's comet is the best-known example of this group, returning every 76 years or so.

Meteors and Meteorites

1. The average meteor—
   a. is about the size of a grain of sand;
   b. moves at about 30 km/sec;
   c. burns up at an altitude of 50 miles.
2. Meteor showers are the debris streams of older comets whose paths cross the Earth's orbit.
3. Meteorites are remnants of asteroids that were large enough to survive passage through the Earth's atmosphere. Their ages, which tell us how long it has been since the Solar System solidified, turn out to be about 4.6 billion years.

Interplanetary Dust and Gas

1. The zodiacal light is produced by the scattering of sunlight from interplanetary dust, distributed mainly in the plane of the Solar System.
2. Radiation pressure from the Sun (Poynting-Robertson effect) forces dust particles to spiral inward on time-scales of a few tens of thousands of years. The dust might be replenished by the erosion of comets, or more probably by collisional grinding of asteroids.
3. Interplanetary gas, flowing away from the Sun, is called the "solar wind." Charged particles in this "wind" push against the magnetic fields of the Earth and Jupiter, creating magnetic tails on the side away from the Sun.

Basic Terms Introduced in This Chapter

Asteroid
Comet
Coma
Solar wind
Meteor
Meteorite
Zodiacal light

TOPICS FOR DISCUSSION AND CLASS ACTIVITIES

1. There was quite a bit of worry among N.A.S.A. scientists as the Pioneer 10 spacecraft, on its way to Jupiter, entered the asteroid belt. True to its name, Pioneer was the first manmade object to venture beyond the orbit of Mars and, if it had found the asteroid zone uncrossable, this would have been a major blow to the space program. Students who want to look into the story more closely should consult the references given for Jupiter in Chapter 9. (See also P. Millman's article, "The Meteoritic Hazard of Interplanetary Travel," *American Scientist,* Nov./Dec. 1971.)

2. The asteroid Eros was once an essential link in the chain of

research that established the distance scale within our Solar System. When Eros came within 22 million kilometers of the Earth in 1931, an international program was undertaken to determine its distance accurately. The details of this project might lead to a good paper for a serious student.

3. Meteorites could, of course, fall anywhere on Earth at any time. There are no verified stories of a meteorite killing anyone, although ancient and recent meteorite falls make for some interesting stories and speculation. Students may want to look into these. In addition, many laypeople as well as scientists enjoy searching out meteor craters all over the world. (See the "Reading Materials," below.)

4. One of the most controversial "falls" occurred near Tunguska in Siberia in 1908. The extent of the destruction has led to speculation that the cause was not a meteor at all. Comets, chunks of antimatter, and "mini-black-holes" have each been suggested as agents of the enormous fireball and subsequent shock wave. However, no real evidence seems available to support these unorthodox explanations.

5. Very rarely a meteor will enter the Earth's atmosphere in a path that carries it back out again. A recent example of such a fireball happened when a tourist couple was taking some films; the resulting (rather spectacular) photographs were published in *Sky and Telescope,* July 1974.

6. As Figure 10.10 in the text indicates, comets have historically been interpreted as omens of major change on Earth. In Shakespears's *Julius Caesar,* for example, Calpurnia comments:

   "When beggars die, there are no comets seen;
   The heavens themselves blaze forth the death of princes."

   Students majoring in literature or history may enjoy searching out other references to comets, and might investigate how the views of comets fit into the cosmology of earlier peoples.

7. Few events since creation have received the sort of publicity Comet Kohoutek got in the winter of 1973-74. Everyone, from Snoopy in the *Peanuts* comic strip to news commentators who had made little distinction between horoscopes and telescopes before this time, was suddenly interested in comets and astronomy. (The interest, we should add, passed more quickly than the comet, particularly when early estimates of the brightness turned out to be over-optimistic.) An analysis of the "Kohoutek months" would make a nice paper. What did astronomers do well or poorly? What can we learn from this episode about the interaction of scientists with modern mass media? Why did so many people get interested, and where did most people get their information? It might also be instructive to compare some of the nonsense engendered by Kohoutek with the stories told of "end-of-the-world" parties during the last pass of Halley's comet.

   And, of course, students may want to do a paper on the *astronomical* things we learned from the comet. In particular, the discovery of several new molecules in Comet Kohoutek seems to shed further light on the early evolution of the Solar System.

8. The Murchison meteorite, which fell on October 28, 1969, in Australia, contained among other substances amino acids whose polarization properties indicate them to be of nonbiological (and nonterrestrial) origin. Biochemists speculate that these organic molecules may be telling us that early steps in the chemical syntheses that led to life on Earth may have occurred elsewhere. (See J. Lawless et al., "Organic Matter in Meteorites," *Scientific American*, June 1972. See also Chapters 11 and 19 of the text.)

9. One enjoyable activity with a class that can meet at night is to schedule an observing session during a meteor shower. These are listed in many books. A free astronomical-events calendar that lists meteor showers and many other useful facts can be obtained by sending a friendly note and a stamped self-addressed envelope to Optica b/c Company, 4100 MacArthur Blvd., Oakland, California 94619.

READING MATERIALS FOR THE STUDENT

General Books

Hawkins, G. *Meteors, Comets and Meteorites* (McGraw-Hill paperback, 1964).

In addition, most of the "General Books" listed for Chapter 9 have chapters on asteroids, comets, and meteors.

Asteroids

Chapman, C. "The Nature of Asteroids," *Scientific American*, January 1975.

Chapman, C., and D. Morrison. "The Minor Planets," *Sky and Telescope*, February 1974.

Duncombe, R., et al. "Toro: The Imprisoned Bull?" *Sky and Telescope*, June 1974.

Hartmann, W. "The Smaller Bodies of the Solar System," *Scientific American*, September 1975.

Lecar, M. "The Asteroids," in *Man and Cosmos*, ed. J. Cornell and E. Hayes (Norton, 1975).

Matthews, M. "The Asteroid Conference in Tucson," *Sky and Telescope*, July 1971.

Meeus, J. "Eros' Closest Approach to the Earth," *Sky and Telescope*, October 1974.

Richardson, R. "The Discovery of Icarus," *Scientific American*, April 1965. [Reprinted in *New Frontiers in Astronomy*, ed. O. Gingerich, W. H. Freeman and Co., 1975.]

Thomsen, D. "Toro! Toro! Toro!" *Science News*, September 15, 1973.

Zallner, B. "New Findings About Eros," *Sky and Telescope*, December 1975.

Comets

Deutschmann, W. "Comets," in *McGraw-Hill Yearbook of Science and Technology*, (McGraw-Hill, 1975).

Jacchia, L. "The Brightness of Comets," *Sky and Telescope*, April 1974.

Maran, S. "A Funny Thing Happened to Comet Kohoutek," *Natural History*,

March 1974.
Maran, S. "Far Out Comet Watching," *Natural History,* October 1974.
Maran, S. "A Cosmic Laboratory," in *Science Year 1975* (World Book Encyclopedia).
Rahe, J., and B. Donn. "The Visual Observation of Comets," *Sky and Telescope,* April 1971.
Roosen, R., and B. Marsden. "Observing Prospects for Halley's Comet," *Sky and Telescope,* June 1975.
Sekanina, Z. "The Prediction of Anomalous Tails of Comets," *Sky and Telescope,* June 1974.
Weaver, K. "What You Didn't See in Kohoutek," *National Geographic,* August 1974.
Whipple, F. "The Nature of Comets," *Scientific American,* February 1974. [Reprinted in *New Frontiers in Astronomy,* ed. O. Gingerich, W. H. Freeman and Co., 1975.]

Meteors and Meteorites

Anders, E. "Diamonds in Meteorites," *Scientific American,* October 1965.
"Australia's Henbury Craters," *Sky and Telescope,* May 1975.
Cameron, I. "Meteorites and Cosmic Radiation," *Scientific American,* July 1973.
Futrell, D. "Some Notes on Tektites," *Sky and Telescope,* May 1967.
Grossman, L. "The Most Primitive Objects in the Solar System," *Scientific American,* February 1975.
Hartmann, W. "The Smaller Bodies of the Solar System," *Scientific American,* September 1975.
Hodge, P. "Central Australia's Meteorite Craters," *Astronomical Society of the Pacific Leaflet* #487 (January 1970).
Jacchia, L. "A Meteorite That Missed the Earth," *Sky and Telescope*, July 1974.
Lawless, J., et al. "Organic Matter in Meteorites," *Scientific American,* June 1972.
Ley, W. *The Meteorite Craters* (Weybright and Talley, 1968).
Mason, B. "Organic Matter from Space," *Scientific American,* March 1963.
O'Keefe, J. "Tektites and Impact Fragments from the Moon," *Scientific American,* February 1964.
Petterson, H. "Cosmic Spherules and Meteoritic Dust," *Scientific American,* February 1960.
Revelle, D. "Studies of Sounds from Meteors," *Sky and Telescope,* February 1975.
Wood, J. "Chondrites and Chondrules," *Scientific American,* October 1963.

READING MATERIALS FOR THE INSTRUCTOR

Anders, E. "Meteorites and the Early Solar System," in *Annual Reviews of Astronomy and Astrophysics,* vol. 9 (1971).
Anders, E. "Extinct Superheavy Element in the Allende Meteorite," *Science,* December 26, 1975.
Gary, G., (Ed.). *Comet Kohoutek* (N.A.S.A. Publication SP-355, 1975). [Forty-two technical papers from a June 1974 conference, summarizing what had been learned from the most famous comet in history.]
Hartmann, W. *Moons and Planets* (Bogden and Quigley, 1972). [See Chapters 7, 8, and 9.]
Marsden, B. "Comets," in *Annual Reviews of Astronomy and Astrophysics,*

vol. 12 (1974).
Mason, B. *Meteorites* (Wiley, 1962).
Middlehurst, B., and G. Kuiper (eds.). *The Moon, Meteorites and Comets* (Univ. of Chicago Press, 1963).
Millman, P. "The Meteoritic Hazard of Interplanetary Travel," *American Scientist,* Nov./Dec. 1971.
Tatum, J. "The Observability of Comets," *Journal of the Royal Astronomical Society of Canada,* October 1975.
Wood, J. *Meteorites and the Origin of Planets* (McGraw-Hill, 1968).

AUDIO-VISUAL AIDS

Films

Comets, Meteors and Planetoids (1963, color, 11 min). Rather routinely done, but informative; good for younger audience. (Distributed by Modern Film Rentals.)

Exploring in Space: The Van Allen Radiation Belts (1963, color, 17 min). Includes a discussion of the radiation belts and the solar wind. (Encyclopaedia Britannica Film.)

Fields of Space (1971, color, 19 min). Introduction to matter and fields between stars. (Made by National Film Board of Canada; distributed by McGraw-Hill Films.)

Interplanetary Space (1969, color, 23 min). Focuses on the solar wind, but also mentions asteroids, comets, and meteors. (Part of "The Sun and the Earth Series," produced under the auspices of the National Academy of Science; distributed by McGraw-Hill Films.)

Minor Members of the Solar System (1959, black and white, 29 min). James Pickering is featured in a discussion of asteroids, comets, and meteors. (Produced by National Educational Television; distributed by Indiana University Audio-Visual Center.)

The Sky Is Falling (1967, color, 24 min). Part of the series on the Smithsonian Institution, this film focuses on work concerning meteors and meteorites. (Produced by NBC News; distributed by McGraw-Hill Films.)

Film Loops

Comet Orbits (1964, color, 3 min). Animation and photos. (Produced by Film Associates of California; distributed by Bailey Film Associates.)

Meteors and Meteorites (1969, color, 4 min). Animation and photos. (Produced by Davidson Films; distributed by Bailey Film Associates.)

Slides

Most observatory slide catalogues include a number of good comet slides. Norton Scientific Company markets a series of 32 slides of meteorites and craters.

Audio Tape

Intruders from Outer Space (36 min). Dr. B. Mason discusses meteors and meteorites. (Distributed by Center for Cassette Studies.)

SAMPLE EXAM QUESTIONS (MULTIPLE-CHOICE)

1. What was the development that made possible a flood of asteroid discoveries in the nineteenth century?
   a. radar
   b. photography
   c. the discovery of Jupiter
   d. spectroscope
   e. the discovery of Eros
2. The group of asteroids whose orbits have been locked into a stable configuration with the same period as Jupiter are given the name--
   a. Icarus
   b. Hidalgo
   c. Trojan
   d. Jovian
   e. Toro
3. The strongest evidence that comets are members of our Solar System comes from—
   a. the shape of their orbits
   b. the shape and length of their tails
   c. the composition of their comas
   d. their association with the planet Pluto
   e. their effects on the lives of important kings and leaders throughout history
4. "Shooting stars" are—
   a. fast-moving stars relatively near to our own
   b. small comets, moving near the Earth
   c. optical illusions
   d. small meteors vaporizing in the Earth's atmosphere
   e. small asteroids that look like stars because they are relatively far away
   f. Burt Reynolds and Robert Redford on a hunting trip
5. Meteor showers are caused by—
   a. the coma of a comet intersecting the Earth's atmosphere
   b. unstable weather conditions over observatories
   c. the Earth moving into the orbit of the debris from an older comet
   d. larger rocks not vaporizing in the Earth's atmosphere and causing disturbances in the clouds
   e. the Earth passing through the asteroid belt
6. As measured by radioactive dating, the age of meteorites turns out to be about—
   a. the time that has passed since the last large comet came near the Earth
   b. a few centuries
   c. 2 million years
   d. 5 billion years
   e. The meteorites, like the Sun, appear to be unmeasurably old.
7. The zodiacal light is produced by—
   a. reflection from the tails of many comets
   b. meteors, as they burn up in the Earth's atmosphere
   c. the solar wind, hitting the upper layers of the Earth's atmosphere
   d. the interplanetary dust, which scatters sunlight
   e. the world's astrologers, as they stay up late into the night casting horoscopes

8. The interplanetary gas comes from—
   a. comets evaporated by sunlight
   b. the outer atmosphere of the Sun itself
   c. the atmospheres of the planets
   d. asteroids that are ground up in collisions
   e. none of the above

Answers to Multiple-choice Questions

1. b 2. c 3. a 4. d 5. c 6. d 7. d 8. b

SAMPLE EXAM QUESTIONS (ESSAY)

1. Briefly discuss the ways in which the gravitational pull of Jupiter has influenced the asteroids. Give some examples.
2. A friend of yours, who lives next door to an astronomer, is given the opportunity to spend a night in the observatory with him, taking many photographs. He returns the next day and tells you that on one of the photographs they found the trail of an asteroid. Are you surprised and/or impressed? Explain your answer.
3. Why are most asteroids irregular in shape?
4. Explain how comets acquire tails. Why do some have two kinds of tails?
5. What is the difference between meteors and meteorites? Name two things we can learn from each of these kinds of cosmic debris.
6. The Earth is not expected to fall into the Sun. Why then do astronomers believe that many of the dust particles in the inner Solar System eventually fall into the Sun?
7. List three effects of the solar wind in our Solar System.

CHAPTER 11
AGES, ORIGINS, AND LIFE

NOTE BY AUTHOR OF TEXT

This is an expansive chapter, because it moves outside hard science into areas that require broad views and judgments. I have tried to distinguish carefully, however. On the question of the origin of the Solar System, I have been conservative and cautious, because this is a problem where there should eventually be a sound theory, although none yet exists. But on the question of life elsewhere in the Universe, where nothing can yet go beyond surmise, I have tried to move from the few facts to the best estimates that can be made.

—*I.R.K.*

TEACHING NOTES

This chapter summarizes the general properties of the Solar System and explores some of the current thinking about the prospects for extraterrestrial life. This latter area is one of tremendous interest among students raised on science-fiction films and stories. It is particularly effective to contrast the scientific approach to this question with the wild and baseless speculation of some popular writers on the subject.

SUMMARY OF KEY POINTS

The Age and Origin of the Solar System

1. Radioactive atoms can be used as clocks that are independent of external conditions.
2. Rocks near the surface of the Earth have been churned and remelted so often that we must look to extraterrestrial materials (Moon rocks and meteorites) to find the age of the Solar System. This turns out to be about 4.6 billion years.
3. General properties of the Solar System:
   a. most of the mass is in the Sun;
   b. most of the angular momentum is in the orbital revolution of the Jovian planets.
4. Stars form from very large clouds of gas and dust. If there is any initial rotation, then as the cloud contracts, the conservation of angular momentum requires the spin to increase to such a degree that the star should become rotationally unstable. Angular momentum can be accomodated in the system—

a. if the system splits into several stars that have orbital motion; or
b. if the angular momentum is transferred to planets.

In case "b," as protostar clouds contract, at first a cool "solar nebula" forms. The central condensation drags against the outer layers, transferring angular momentum.

5. As the Solar System formed, material coalesced into smaller condensations, which collided and merged to form the planets. Different elements condensed out of the gas at different temperatures, thus accounting for some of the composition differences between the inner and outer planets.

Life in the Universe

1. Only the carbon atom has the ability to form the many different combinative patterns that living organisms require; thus life elsewhere will probably be *chemically* similar to us.
2. Life on Earth has developed in a water environment. A liquid seems necessary to help reagents come into contact. The speed of the reactions that then take place is a function of temperature. Other common substances liquefy at prohibitively low temperatures.
3. Only a narrow temperature range permits life. If the temperature is too high, molecules break up; if too low, reactions go too slowly.
4. Thus planetary surfaces would be the best environments in which to search for life.
5. In our own Solar System, Mars may have had life during more favorable times. The atmospheres of Jupiter and the other giant planets may have layers warm enough for life to form.
6. Beyond our Solar System, we estimate that there may be a billion or more planets suitable for life.
7. Evidence for relatively easy chemical origin of life:
   a. synthesis of many organic compounds in the laboratory;
   b. molecules discovered in interstellar clouds;
   c. molecules in meteorites.
8. If other intelligent civilizations exist, radio waves represent an efficient way of contacting them.

Basic Terms Introduced in This Chapter

Half-life
Angular momentum
Conserved quantities

TOPIC FOR DISCUSSION AND CLASS ACTIVITIES

1. Students of the history of ideas may recognize in the two main kinds of theories concerning the origin of the Solar System discussed in the text the familiar contrast between *catastrophist* and *uniformitarian* notions. As in geology and biology, early proposals that some sort of unique catastrophe was needed to bring about present-day conditions on Earth have now been replaced with models that rely on the action of steady, generally uniform processes. Students in other fields of science may want to do papers comparing this trend in astronomy with the development of theories in other fields. (See for example Loren Eiseley, *Darwin's Century*, Doubleday Anchor paperback, 1958.)

2. In thinking about the search for life on other planets, it is interesting to consider how hard it would be for extraterrestrial astronomers or explorers to detect life (and particularly intelligent life) on Earth. Carl Sagan, of Cornell University, has written several nice pieces on this subject. (See for example Chapter 18 of *Intelligent Life in the Universe*, by I. Shklovskii and C. Sagan, Delta paperback, 1966.)

3. It is interesting to speculate what forms life would take in different environments—for example, the clouds or surface of Venus, the Martian desert, or the Jovian atmosphere. Some of this kind of speculation is well described in Carl Sagan's more recent, popular-level book, *The Cosmic Connection* (Dell paperback, 1973).

4. Beyond the Apollo mission samples (which indicated that the Moon is sterile), the first *in situ* search for extraterrestrial life is the Viking mission, which landed on Mars during the summer of 1976. The results of the Viking studies will be found in various journals for late 1976 and 1977. The following articles, written before the mission, describe the biology experiments.

   Eberhart, J. "The Road to Mars," *Science News*, August 2, 1975.
   Klein, D. "On the Viking Search for Life on Mars," *Mercury*, Jan./Feb. 1975.
   Palsson, R. "The Challenge of Viking: Is There Life on Mars?" *Mercury*, March/April 1976.
   Sagan, C. "Viking to Mars: The Mission Strategy," *Sky and Telescope*, July 1975. [Also see the August and September 1975 issues.]

5. Although the text briefly refers to his work, Peter van de Kamp is not mentioned by name. Whatever the ultimate decision on his observations of dark (perhaps planetary) companions around nearby stars, the research of this Swarthmore astronomer is a fine case study in the need for patience in astronomical projects. Since the year 1938, van de Kamp and his coworkers have used each clear night at the Sproul Observatory to photograph the position of the nearest stars, thus following their paths through our sky over the decades. See for example P. van de Kamp, "Barnard's Star: The Search for Other Solar Systems," *Natural History*, April 1970. See also the film *Invisible Planet* ("Audio-visual Aids," below).

6. The text briefly mentions the double bind of space travel: Given the enormous distances between stars, if you travel slowly you won't live to enjoy your arrival; and if you travel quickly you can't afford the trip. Students, particularly those who are devoted to science-fiction stories such as *Star Trek*, will often find this an interesting area for further investigation. Many of the books listed under "Life in the Universe" in the "Reading Materials for the Student" below contain chapters on this topic. See also:

   Purcell, E., et al. "Space Travel: Problems of Physics and Engineering," in *Project Physics Course Reader*, vol. 5 (Holt, Rinehart and Winston, 1971), p. 197.
   Von Hoerner, S. "The General Limits of Space Travel," in *Interstellar Communication*, ed. A. Cameron (Benjamin paperback, 1963).

7. Despite the improbability of space travel, the notion of a trip through space continues to have a strong hold on the public imagination. Recently there has been a tremendous resurgence in UFO sightings, with the usual mindless media coverage. It seem clear that the widespread desire to make immediate contact with advanced extraterrestrials is at least partly the result of age-old religious needs that are apparently not being met for many people in modern society. Any valid analysis of the UFO phenomenon must take this psychological side of the question into consideration. The following are among the few responsible books on UFOs.

Condon, E., et al. *Scientific Study of UFO's* (Bantam paperback, 1969). [A huge volume, detailing the findings of a scientific panel, funded by the Air Force, which concluded basically that UFOs are *not* extraterrestrial and are not worth studying much further.]

Hynek, J. *The UFO Experience: A Scientific Inquiry* (Regenery, 1972). [Perhaps the best-known astronomer who leans toward the extraterrestrial hypothesis, Hynek outlines his opposition to the conclusions of the Condon report.]

Jacobs, K. *The UFO Controversy in America* (Indiana Univ. Press, 1975). [Among other things, this is a sober study of the studies, focusing on the story of the Air Force panel. The author, however, tends toward the "unexplained" viewpoint for a number of sightings.]

Klass, P. *UFO's Explained* (Random House, 1975). [Carefully documented discussion of the various natural and manmade phenomena that can be mistaken for flying saucers and the like.]

Sagan, C., and T. Page (eds.). *UFO's: A Scientific Debate* (Norton paperback, 1972). [Papers from a special symposium held by the A.A.A.S. to "clear the air" following the furor that erupted over the Air Force study of UFOs. All sides were given time to put their best evidence forward.]

8. Another example of the public's willingness to accept poorly supported claims in which they have a predisposition to believe is the outrageous success of Erich von Daniken, a convicted embezzler and forger, who claims that ancient astronauts visited the Earth a long time ago and affected our cultural development. Here are some objective references to which interested students can be referred:

Bracewell, R. "The Chariots of von Daniken," in *The Galactic Club* (W. H. Freeman and Co. paperback, 1974), Chapter 11.

Krupp, E. "The von Daniken Phenomenon," *Griffith Observer*, April 1974.

Lingeman, R. "Erich von Daniken's Genesis," *The New York Times Book Review*, March 31, 1974.

McIntyre, L. "Mystery of the Ancient Nazca Lines," *National Geographic*, May 1975.

Story, R. *The Space Gods Revealed: A Closer Look at the Theories of E. von Daniken* (Harper & Row, 1976). [Foreword by Carl Sagan.]

In May 1974, the Division of Interdisciplinary and General Studies at the University of California, Berkeley, sponsored a symposium concerning the works of Velikovsky, von Daniken, and Donnelly (who wrote on Atlantis). Entitled "Lost Worlds and Golden

Ages," the symposium brought together scholars from several fields who examined the ideas of the three men critically and unanimously found them wanting (although there was mild agreement that they were good yarns). Tapes of the talks are available from the Extension Media Center, Univ. of Calif., Berkeley, Calif. 94720.

9. Several crude attempts to search for radio signals from extraterrestrial civilizations have already been made by Soviet and American astronomers. (See, for example, the article by Carl Sagan and Frank Drake in *Scientific American,* May 1975.)

   When the newly resurfaced 1,000-foot radio telescope at Arecibo, Puerto Rico was rededicated on November 16, 1974, the ceremony included the first broadcast of a radio message *from* Earth, beamed in the direction of the globular cluster M13. (See, e.g., *Science News,* November 23, 1974.) Students often enjoy composing and discussing such messages, and then exploring the anthropocentrism of much of our thinking about interstellar communication.

10. If we ever do make contact with extraterrestrial intelligence, what will be the consequences for science, religion or philosophy, racism, psychology, etc.? These prospects can spark lively class debate. Later, views could be compared with those of a panel of experts brought together by Boston University and N.A.S.A. in 1972 to consider such questions. Chaired by astronomer/historian Richard Berendzen, the panel included anthropologist Ashley Montagu, physicist Philip Morrison, astronomer Carl Sagan, theologian Krister Stendahl, and biologist George Wald. Their comments are transcribed in a book, *Life Beyond the Earth and the Mind of Man,* available from the U.S. Government Printing Office as N.A.S.A. publication SP-328. An edited film (with the same name) is available from N.A.S.A.

11. Implicit in the discussion of the text is the belief, shared by most scientists, that life on Earth came about as a result of a series of slow chemical (and ultimately biological) processes. For many religious people, this notion is as disturbing as the concept of biological evolution. Recently, fundamentalists and others have renewed efforts to change gradeschool textbooks that espouse only the scientific point of view. For an interesting analysis, students (and teachers) should read "The Science Textbook Controversy," by D. Nelkin, *Scientific American,* April 1976.

12. Another unusual approach to the origin-of-life problem is to postpone its consideration by proposing that life originated elsewhere and then came to the Earth in some protected form. A version of this theory, called panspermia, was advanced by the Swedish chemist Arrhenius in 1907. In his view, life spores traveled from star to star, propelled by radiation pressure. This notion is analyzed carefully in Chapter 15 of *Intelligent Life in the Universe,* by Shklovskii and Sagan.

    More modern approaches to this question include the possibility of "directed panspermia"—i.e., the seeding of other solar systems by intelligent civilizations. (See an article by Crick and Orgel in *Icarus,* July 1973.) Some writers have even considered "accidental panspermia" hypotheses, such as the suggestion by Thomas Gold that we are all descended from the refuse left here by some careless interstellar picnickers.

13. Students who would like a scientifically plausible science-fiction story about interstellar contact can be referred to two interesting novels: *The Black Cloud,* by Fred Hoyle, and *The Listeners,* by James Gunn. Also, an interesting example of a life-form based on silicon can be found in the early short science-fiction story "A Martian Odyssey," by Stanley Weinbaum. It is just one of a number of stories based on more-or-less plausible science, reprinted with commentary by Isaac Asimov in *Where Do We Go from Here?* (Fawcett paperback, 1971).

READING MATERIALS FOR THE STUDENT

The Age and Origin of the Solar System

A number of the books listed for Chapter 9 have sections devoted to this subject.

Cameron, A. "The Origin and Evolution of the Solar System," *Scientific American,* September 1975.

Grossman, L. "Condensation of the Solar System," in *McGraw-Hill Yearbook of Science and Technology* (McGraw-Hill, 1974).

Huang, S. "Disk and Ring Structure in the Universe," *Sky and Telescope,* April 1972.

Lewis, J. "The Chemistry of the Solar System," *Scientific American,* March 1974.

Lyttleton, R. "The Origin of the Solar System," in *Astronomical Society of the Pacific Leaflets* #471 and #472 (September, October 1968).

Reynolds, J. "The Age of the Elements in the Solar System," *Scientific American,* November 1960.

Life in the Universe: General Books

Berendzen, R. *Life Beyond the Earth and the Mind of Man* (N.A.S.A. Publication SP-328, 1973). [See Topic #10, above.]

Bracewell, R. *The Galactic Club* (W. H. Freeman and Co. paperback, 1974).

Dole, S. *Habitable Planets for Man,* 2nd ed. (Elsevier, 1970).

Jastrow, R. *Red Giants and White Dwarfs* (Signet paperback, 1968). [Elementary primer on the evolution of the cosmos, from galaxies to life.]

Maruyama, M., and A. Harkins (eds.). *Cultures Beyond the Earth* (Vintage paperback, 1975). [A symposium of the American Anthropological Association.]

Sagan, C. *The Cosmic Connection* (Dell paperback, 1973).

Shklovskii, I., and C. Sagan. *Intelligent Life in the Universe* (Delta paperback, 1966).

Sneath, P. *Planets and Life* (Funk and Wagnalls, 1970).

Sullivan, W. *We Are Not Alone,* 2nd ed. (Signet paperback, 1966).

Life in the Universe: General Articles

Hart, M. "An Explanation for the Absence of Extra-terrestrials on Earth," *Quarterly Journal of the Royal Astronomical Society,* June 1975.

Huang, S. "Life Outside the Solar System," *Scientific American,* April 1960.

Sagan, C., and F. Drake. "The Search for Extra-terrestrial Intelligence," *Scientific American*, May 1975.
Thomsen, D. "Extra-terrestrial Life (If There Is Any) and Us," *Science News*, July 14, 1973.
Thomsen, D. "Tuning in on Other Worlds," *Science News*, November 15, 1975.
van den Bergh, S. "Are We Alone?" *Mercury*, July/August 1975.

Origin and Development of Life on Earth
Barber, V. "Theories of the Chemical Origin of Life on the Earth," *Mercury*, Sept./Oct. 1974.
Benford, G. "The Hospitable Planet Earth," *Natural History*, December 1973.
Fox, S., and R. McCauley. "Could Life Originate Now?" *Natural History*, Aug./Sept. 1968.
Kiefer, I. "Chemical Evolution," *Smithsonian*, May 1972.
Miller, S. "The First Laboratory Synthesis of Organic Compounds Under Primitive Earth Conditions," in *The Heritage of Copernicus*, ed. J. Neyman (MIT Press, 1974).

Evidence for Life Out There
Buhl, D. "Light Molecules and Dark Clouds," *Mercury*, Sept./Oct. and Nov./Dec. 1972.
Buhl, D. "Molecules and Evolution in the Galaxy," *Sky and Telescope*, March 1973.
Driscoll, E. "A Planet for the Sun's Twin," *Science News*, February 10, 1973.
"How Common Are Planetary Systems?" *Sky and Telescope*, July 1975.
Jastrow, R. "Cosmic Evolution," *Natural History*, January 1968.
Kumar, S. "Planetary Systems," in *The Emerging Universe*, ed. W. Saslaw and K. Jacobs (Univ. Press of Virginia, 1972).
Lawless, J., et al. "Organic Matter in Meteorites," *Scientific American*, June 1972.
Mason, B. "Organic Matter from Space," *Scientific American*, March 1963.
McCray, R. "Molecules Between the Stars," *Natural History*, December 1974.
Ney, E. "The Mysterious Egg Nebula," *Sky and Telescope*, January 1975.
Turner, B. "Interstellar Molecules," *Scientific American*, March 1973.
van de Kamp, P. "Barnard's Star: The Search for Other Solar Systems," *Natural History*, April 1970. [But see also new evidence, for example in the "Noted in the Current Journals" column in *Mercury*, Jan./Feb. 1974.]

See also the articles about the Viking mission to Mars, cited under Topic #4 above.

Miscellaneous
Bracewell, D. "Interstellar Messengers," *Mercury*, March/April 1975.
Dalzell, B. "Exotic Bestiary for Vicarious Space Voyagers," *Smithsonian*, October 1974.
Duckworth, E. "In Search of the Galactic Library," *Mercury*, May/June 1975.
Stern, D. "First Contact with Non-human Cultures," *Mercury*, Sept./Oct. 1975.

READING MATERIAL FOR THE INSTRUCTOR

The Age and Origin of the Solar System

Alfven, H. "Plasma Physics, Space Research and the Origin of the Solar System," *Science*, June 4, 1971.

Ward, W. "Formation of the Solar System," in *Frontiers of Astrophysics*, ed. E. Avrett (Harvard Univ. Press, 1976).

See also references for Chapter 9.

Life in the Universe: General Books

Mamikunian, G., and M. Briggs (eds.). *Current Aspects of Exobiology* (Pergamon Press, 1965).

National Aeronautics and Space Administration. *Project Cyclops Report* (N.A.S.A., 1971). [Available from Dr. John Billingham, N.A.S.A. Ames Research Center, Code LT, Moffett Field, Calif. 94035.]

Ponnamperuma, C., and A. Cameron (eds.). *Interstellar Communication* (Houghton-Mifflin paperback, 1974).

Sagan, C., (ed.). *Communication with Extra-terrestrial Intelligence* (MIT Press, 1973).

Life in the Universe: Articles

Anders, E., et al. "Organic Compounds in Meteorites," *Science*, November 23, 1973.

Buhl, D., and C. Ponnamperuma. "Interstellar Molecules and the Origin of Life," in *Space Life Sciences*, vol. 3 (1971), p. 157.

Calvin, M. "Chemical Evolution," *American Scientist*, March/April 1975.

Huang, S. "Occurrence of Planetary Systems in the Universe as a Problem in Stellar Astronomy," in *Vistas in Astronomy*, vol. 11, ed. A. Beer (Pergamon Press, 1969).

Kahn, F. "Life in the Universe," in *The Emerging Universe*, ed. W. Saslaw and K. Jacobs (Univ. Press of Virginia, 1972).

Oliver, B. "Life in the Galaxy," *Journal of College Science Teaching*, December 1973.

Oliver, B. "Proximity of Galactic Civilizations," *Icarus*, vol. 25, p. 360 (1975).

Prigogine, I., et al. "Thermodynamics of Evolution," *Physics Today*, Nov.-Dec. 1972.

Snyder, L. "Chemistry Between the Stars," in *The Emerging Universe*, ed. W. Saslaw and K. Jacobs (Univ. Press of Virginia, 1972).

Solomon, P. "Interstellar Molecules," *Physics Today*, March 1973.

van de Kamp, P. "Unseen Astrometric Companions of Stars," in *Annual Reviews of Astronomy and Astrophysics*, vol. 13 (1975).

Zuckerman, B., and P. Palmer. "Radio Radiation from Interstellar Molecules," in *Annual Reviews of Astronomy and Astrophysics*, vol. 12 (1974).

AUDIO-VISUAL AIDS

Films

Evolution and Origin of Life (1970s, color, 33 min). A nice introduction to this field; a good way to give astronomy students some biology background. (CRM Educational Film.)

Invisible Planet (1966, black and white, 30 min). A sentimental but well-done look at Peter van de Kamp and his possible discovery of a dark companion around Barnard's Star. (Part of the "Experiment" series, distributed by Indiana University.)

The Ladder of Creation (1974, color, 52 min). Jacob Bronowski reviews the work of Wallace and Darwin on evolution. Includes a brief discussion of experiments in prebiotic synthesis and of interstellar molecules. (Part of the "Ascent of Man" series; distributed by Time-Life Films.)

Life Beyond the Earth and the Mind of Man (1975, color, 28 min). Part of a symposium featuring Richard Berendzen, Carl Sagan, Ashley Montagu, Phillip Morrison, George Wald, and Krister Stendahl discussing extraterrestrial life from many points of view. (N.A.S.A. film HQ-245.)

Planetary Biology at Ames (1970s, color, 22 min). A good, if somewhat over-produced and condescendingly narrated, summary of (1) N.A.S.A. laboratory work on the synthesis of organic substances, (2) the discovery of organic materials in meteorites, (3) interstellar molecules, and (4) the Viking search for life on Mars. (N.A.S.A. film NAV-029.)

Search for Life (1975, color). This superb new film is perhaps the best astronomically oriented look at questions about extraterrestrial life. A number of astronomers and biologists are featured in nicely edited sequence. Should be available for national distribution soon. (Part of the "Nova" series on public television.)

Space: Life Out There (1973, color, 22 min). Carl Sagan and others consider extraterrestrial life and space exploration. (Part of the "Towards the Year 2000" series; distributed by Document Associates.)

To Find Life on Other Planets: Are the Martians Waiting for Us? (1974, color, 22 min). Mainly about the Viking mission, this film reviews past and planned experiments in exobiology. (Part of the "Target: The Impossible" series; distributed by Document Associates.)

We Are Not Alone: Is There Life on Other Planets? (1967, black and white, 31 min). Based on the prize-winning book by Walter Sullivan, this film reviews the experiments and speculation concerning life out there. (Distributed by McGraw-Hill Films.)

Who's Out There? (1975, color, 28 min). Narrated by Orson Welles, this carefully planned and effective film is an excellent introduction to the scientific study and social implications of the search for extraterrestrial life. (N.A.S.A. film HQ-226.)

Audio Tapes

As It Was in the Beginning (26 min). Dr. Robert Jastrow discusses the evolution of the Solar System. (Distributed by Center for Cassette Studies.)

Intelligent Life in the Universe (20 min). Interview with Dr. Ronald Bracewell, Stanford astronomer and author of *The Galactic Club*. (Astronomical Society of the Pacific cassette ER-143.)

Is Anybody Out There? (43 min). Dr. Bernard Oliver, one of the prime movers behind the Cyclops study, discusses extraterrestrial life and communication. (Distributed by Center for Cassette Studies.)

The Origins of Life (39 min). Dr. William Schopf discusses how life developed on the Earth. (Distributed by Center for Cassette Studies).

The Search for Life on Mars (20 min). Interview with Dr. Harold Klein, Chief Project Scientist of the Viking Mission. (Astronomical Society of the Pacific cassette ER-136.)

UFO's (2 cassettes, 30 min each). Interviews with Philip Klass and Roy Pritchard, with terrestrial explanations for UFOs. (Center for Cassette Studies.)

SAMPLE EXAM QUESTIONS (MULTIPLE-CHOICE)

1. The reason that radioactive atoms are such reliable clocks for dating the Earth is that—
   a. *all* radioactive atoms break up at the same rate
   b. radioactivity occurred only during the first million years or so after the Earth formed
   c. their decay rate does not depend on external circumstances
   d. all radioactive nuclei break up exactly after *half* of their lifetime is over
   e. we can now predict *exactly* when a given radioactive nucleus is going to break up
2. Most of the angular momentum in our Solar System is in—
   a. the Sun
   b. the Jovian planets
   c. the Earth (because it turns so much faster)
   d. Mercury
   e. Pluto
3. Which of the following is *not* a useful argument *against* the hypothesis that planets form by the tidal interaction of two stars that pass near each other?
   a. The material of the Jovian planets seems similar to the Sun's atmosphere.
   b. Near encounters of stars are very rare in our huge Galaxy.
   c. Not enough angular momentum can be transferred in the process.
   d. Material torn out of the hot Sun would dissipate.
   e. *All* the above are good arguments against the hypothesis.
4. According to our modern picture of star formation, which of the following is a way for a new star-cloud to avoid becoming rotationally unstable as it contracts?
   a. split into two or more stars
   b. form a system of planets
   c. have a near encounter with a nearby star, thus slowing down
   d. have a lot of collisions within the cloud
   e. more than one of the above
5. Which chemical element is best for making the wide variety of combinative patterns necessary for complex organic molecules?
   a. water
   b. silicon
   c. carbon
   d. oxygen
   e. DNA
6. Aside from the Earth, which of the following planets would be the most likely place to discover life?
   a. Mercury
   b. Venus
   c. Mars
   d. Pluto
   e. Neptune
7. Biochemists have had considerable success in synthesizing organic molecules from the gases of the early Earth's atmosphere. A good

source of energy for these reactions in the laboratory is—

a. movement of the solution (simulating ocean waves)
b. radio waves of high frequency
c. X-ray (simulating solar radiation)
d. electrical discharges (simulating lightning)
e. sound wave (simulating thunder)

8. Which of the following is *not* a piece of evidence supporting the notion that the origin of life is a relatively easy process?
   a. Russian spacecraft have brought back fossils from Venus.
   b. Experiments have produced organic molecules by sparking electrical discharges through water, methane, and ammonia gases.
   c. Molecules of organic matter have been discovered in interstellar space.
   d. Amino acids have been isolated from meteorites.

Answers to Multiple-choice Questions

1. c 2. b 3. a 4. e 5. c 6. c 7. d 8. a

SAMPLE EXAM QUESTIONS (ESSAY)

1. Compare the information we receive about the age of the Solar System from rocks on Earth, from Moon rocks, and from meteorites.
2. Geologists continue to search for (and to find) older and older rocks on the Earth's surface. Would you expect that they will eventually find rocks 4.6 billion years old? Explain.
3. A friend of yours, who has never had the benefit of an astronomy course, reads in a local newspaper that all stars in our Galaxy are in motion. In a flash of inspiration, it occurs to him that perhaps stars come near to each other as they move around, and that they pull material from each other to form planets. How would you (gently) dissuade him from this notion?
4. Briefly describe the modern theory of Solar System formation. How does this theory account for (a) the distribution of angular momentum in the Solar System; and (b) differences in the composition of the inner and outer planets?
5. Name three different details of our Solar System that astronomers still have trouble fitting into the general scheme of how the planets formed.
6. Many science-fiction stories have extraterrestrial beings whose chemistry is based on silicon rather than carbon. Discuss how believable such aliens would be. Would they look like us?
7. In a science-fiction story, the astronaut hero meets a multi-tentacled monster in the hot ammonia oceans of Mercury. How plausible is this situation? Explain.
8. You write a science-fiction story about intelligent beings living on a planet in another solar system, and take some pride in your ingenuity in imagining plausible life forms quite different from those of Earth. To your dismay, the editor returns your story with a note chiding you for your lack of imagination! "Everyone writes about extraterrestrial beings that live on planets," he complains. "Why don't you get away from these Earth-based assumptions and rewrite your story using life forms that have not evolved on a planetary surface?" Given our present understanding of the formation of life, how plausible do you judge the editor's

suggestion to be? Explain your answer.

9. Many scientists now believe that life may begin with relative ease in many different environments. Cite at least three pieces of evidence supporting this view.

10. A congressional committee has charged you with reporting to them on the most efficient way of finding other congressional committees out in space. What method would you recommend, and why?

CHAPTER 12
INTRODUCING THE STARS

NOTE BY AUTHOR OF TEXT

This chapter is a pivotal one. First, it makes the transition from our little Solar System commune to the stellar Universe. (You may wish to have the students review the hierarchy of astronomical distances on pp. 4-6.) More important, it shows how the astronomer can logically begin the study of the stars. The subhead "The Entering Wedge" is a key phase, and I have always thought it important to emphasize how crucial is this first step of finding out stellar distances. A related point that needs underlining is the central role of the distance problem in astronomy. Although it is difficult to get students to think easily in terms of distance modulus, the photometric measurement of distance comes up so frequently in the remainder of the book that it needs to be established clearly here.

Chapter 12 and 13 form a related unit, and in some sense Chapter 12 is a prelude to a logical climax in Chapter 13. At the same time, many of the fundamental quantities of stellar astronomy are introduced here; and they must be learned for their own sake, as the language in which the later story will be told.

—I.R.K.

TEACHING NOTES

This short chapter treats some of the fundamental observational parameters in stellar astronomy. Some of the concepts may be difficult for nonscience students (for example, the backward logarithmic nature of magnitudes). It helps to emphasize what these measurements are good for and how difficult they often are to make. Students should not lose sight of the immensity of the distances and the minuteness of many of the effects being discussed. Students having trouble might be encouraged to start keeping an index-card glossary of the more important and difficult concepts, or to check on their understanding by explaining the concepts to any members of their families who can be trapped into listening.

Specific examples and a little humor can help a lot during lectures on these subjects.

(The method of statistical parallax, although very important in astronomical research, is very often beyond the ability of most students without mathematical background, at least without a much longer explanation. However, leaving it out would not affect the clarity of the remaining chapters, and many teachers may want to do just that.)

SUMMARY OF KEY POINTS

1. In order to study the stars, we need to know the distances to them. Trigonometric parallax, using the Earth's orbit as a baseline, makes independent distance measurements possible.
2. Stellar brightness is measured in magnitudes:

   $$m_2 - m_1 = 2.5 \log (b_1/b_2).$$

   Apparent magnitude, m, indicates how bright the star appears from Earth. Absolute magnitude, M, is a measure of the star's brightness at a standard distance (10 parsecs). The two are related to the distance, r, by the expression

   $$m - M = 5 \log r - 5.$$

   The quantity $(m - M)$ is called the distance modulus.
3. Stars can be divided into recognizable types (or classes), within which each star has the same absolute magnitude. Once this relationship between type and absolute magnitude is calibrated, a measurement of apparent magnitude and type can yield the distance.
4. Stellar motions:
   a. radial velocity can be determined by observing the Doppler shift in a spectrum;
   b. transverse velocity can be found only by measuring the proper motion and the distance;
   c. proper-motion studies help identify the nearest stars and compare the average motions of groups of stars.
5. The solar motion defined with respect to the local standard of rest can be measured from the average radial velocities of stars in different portions of the sky. Its value is 20 km/sec toward a point near Vega.
6. The methods of statistical parallax use the average motions of groups of stars to get average distances and, if the group consists of stars of the same type, average absolute magnitudes.

Basic Terms Introduced in This Chapter

Trigonometric parallax
Parsec
Apparent magnitude
Absolute magnitude
Distance modulus
Doppler effect
Proper motion
Statistical parallax
Local standard of rest

The method of spectroscopic parallax is introduced in this chapter, although that name is not used in the text.

TOPICS FOR DISCUSSION AND CLASS ACTIVITIES

1. One quick activity to help students understand parallax is to ask them to place a pencil or pen close to their noses and close first one eye and then the other. Relative to the background (e.g., the blackboard or the instructor), the position of the pencil should shift considerably. Now slowly have them bring the pencil out to arm's length, continuing alternately to close each eye. The shift decreases markedly with distance.

2. To explain the Doppler effect, it is useful to employ an analogy. Let students think about a tennis-training machine, which is programmed to throw balls at a frequency of 3 per minute. What happens to the frequency of ball arrival if the machine rolls continuously toward the player or away from him? What happens to the distance between balls?

3. It helps tremendously to introduce a human element into the rather abstract subjects of this chapter. Instructors with experience in observational astronomy would do well to discuss the practical problems involved in making the measurements the text discusses. The difficulties of guiding, the vagaries of the weather, the likelihood of mistakes when one has been awake all night—all serve to emphasize the hard work that goes into each measurement. Students should then consider why it is that astronomers think these observations important enough to continue making them.

4. The astronomer Hipparchus (mentioned on p. 224 of the text) compiled a very careful catalogue of star positions and characteristics in the second century B.C. In 1718, Edmund Halley found that several nearby stars had moved from the positions given by Hipparchus by an angle so large that the discrepancy could not simply be an error in the measurements of the Greek astronomer. This was the first evidence for proper motions, and would make a nice paper or oral report for the historically minded student. (See, e.g., Z. Kopal, *Widening Horizons*, Taplinger, 1970, Chapter 4.)

5. The discussion of solar motion and the local standard of rest can be a good time to talk a little bit about frames of reference and relative motion. Most students can think of good examples with passing trains or automobiles where only *relative* motion is apparent. From here they can be led to asking if any frames of reference are absolute. Even relatively sophisticated students are sometimes amazed when they learn of the enormous speeds with which the Earth is moving through space. (Very rough calculations give the following speeds in familiar units: equatorial rotation, 1,000 miles/hour; orbital motion, 66,000 miles/hour; solar motion, 43,000 miles/hour; revolution about the Galaxy, 600,000 miles/hour.)

6. Self-contained "laboratory" exercises on the Doppler shift of Saturn's rings and the proper motion of 61 Cygni are available from the Sky Publishing Corporation, 49-50-51 Bay State Road, Cambridge, Mass. 02138.

READING MATERIALS AND AUDIO-VISUAL AIDS

All the materials on stellar astronomy are listed in Chapter 13.

SAMPLE EXAM QUESTIONS (MULTIPLE-CHOICE)

1. Why were astronomers at the time of Copernicus unable to measure stellar parallax?
   a. Measurements had to be made from widely separated parts of the Earth; transportation was not good enough.
   b. At that time, the stars had not moved a significant distance from their first recorded positions.
   c. Observing techniques were not precise enough.
   d. Kepler's laws were not understood then, and the measurements could not be understood without those laws.
   e. At that time, the stars were fixed in the solid crystalline sphere of the heavens.
2. I am observing two stars, one of magnitude 6 and the other of magnitude 11. These stars differ in brightness by a factor of—
   a. 5
   b. 17
   c. 10^5
   d. 100
   e. 11/6
3. I observe a previously uninvestigated star and find its apparent magnitude. To determine its absolute magnitude, I need to find its—
   a. distance
   b. color
   c. temperature
   d. brightness as seen from Earth
   e. social security number
4. To find the radial velocity of a star, I need to measure its—
   a. distance
   b. parallax
   c. Doppler shift
   d. proper motion
   e. absolute magnitude
5. To find the transverse velocity of a star, I need to measure its—
   a. proper motion
   b. Doppler shift
   c. distance
   d. solar motion
   e. more than one of the above
6. The easiest way to pick out the *nearby* stars in a large group of stars is to find their—
   a. apparent brightness
   b. proper motion
   c. spectral type
   d. Doppler shift
   e. None of the above can help pick out nearby stars.
7. The method of statistical parallax can tell us—
   a. the average apparent magnitude of any group of stars
   b. the distance to any star in the Galaxy
   c. the distance the Earth travels in its orbit, on the average

d. the average absolute magnitude of a group of stars of the same type
e. the Sun's position in the Galaxy

8. To get the solar motion, astronomers must measure—
a. the parallax of the star Vega
b. the absolute magnitude of groups of stars in many directions
c. the proper motion of a large group of stars all in the same direction from us
d. the average distance to a group of stars that have the same absolute magnitude
e. the radial velocity of groups of stars in many directions

Answers to Multiple-choice Questions

1. c 2. d 3. a 4. c 5. e 6. b 7. d 8. e

SAMPLE EXAM QUESTIONS (ESSAY)

1. You are an astronomer setting out to measure the parallax of a star. Briefly describe what observations you would need to make and when. How would you decide whether your observation is reliable?
2. How would you measure the distances to stars too far away to show a measurable trigonometric parallax?
3. In many science-fiction stories, astronomers are the first to discover alien spaceships approaching the Earth. Suppose such a ship reflects sunlight so effectively that it has an apparent magnitude large enough to be observable. Describe what measurements you would make to determine its motion.
4. Suppose an astronomer discovers that the star in which he is interested is shown in a reliable eighteenth-century catalogue of stars to be in a slightly different position than that recorded in modern measurements. Describe the observing program he should undertake to get the distance and space velocity of this star.
5. How can astronomers use their measurement of the solar motion to get the average distance of a large group of stars?

CHAPTER 13
SORTING OUT THE STARS

NOTE BY AUTHOR OF TEXT

This chapter is in some sense the highpoint of the book. On the factual level, it introduces stellar spectra and the HR diagram; on the conceptual level, it plunges the student directly into stellar ages and the whole problem of evolution; and on the theatrical level, it is the highpoint of the detective-story metaphor. Starting with the simplest facts of spectral classes and absolute magnitudes, it focuses on the problems of fuel supply and the existence of a main-sequence turnoff, and it then deduces the whole scenario of stellar birth and evolution.

--I.R.K.

TEACHING NOTES

Although the concept of spectral types is introduced and used extensively here, stellar spectra per se are not defined and explained until Chapter 16. Some instructors may wish to assign that chapter first, or at least simultaneously.

Many students who lack science and math background have trouble with the abstract thinking the HR diagram requires. Indeed, there are students in each class who will probably interpret motion on the HR diagram to be physical motion in space. (In this connection, all teachers ought to read "Effective Astronomy Teaching: Intellectual Development and Its Implications," by Dennis Schatz and Anton Lawson, *Mercury*, July/August 1976. Possible activities are suggested in "Topics for Discussion," below.)

SUMMARY OF KEY POINTS

1. Hertzsprung and Russell found two characteristics to be useful in classifying stars and studying their evolution:
    a. absolute magnitude;
    b. temperature (as determined from spectra and denoted by a spectral type: OBAFGKM, with subclasses from 0 to 9).
2. When these chatacteristics are displayed graphically (on an HR diagram), stars fall into several regions or sequences. Most stars fall on the "main sequence," running from upper left to lower right.

3. Roughly speaking, the ratio of mass to luminosity determines how long a star can live (keep shining). Very luminous stars "consume" their mass very quickly and can "die" in only a million years or so, while the dimmest stars can last a hundred trillion years.
4. The more luminous (massive) a star, the more quickly it evolves. As time passes, stars from farther and farther down the main sequence will be evolving away from that region.
5. Thus, when we plot the HR diagram of a cluster of stars (in which all members presumably were born together), the "turnoff point" on the main sequence can tell us the age of the cluster.
6. Stars can be divided into two classes according to the abundance of heavier elements (called "metals"):
   a. Population I refers to metal abundances like that of our Sun (2-3%), found for stars in the galactic disk;
   b. Population II refers to low metal abundances (less than 1%), found for stars in the galactic halo--in particular, in globular clusters. Population II stars are found only in old groups, indicating that metal-poor stars are no longer being made in our Galaxy.
7. A true stellar census should not be biased toward the more easily observable stars of high luminosity. Blink comparators enable us to find all the nearby stars by their large proper motion (seen on photographs taken years apart). The luminosity function is a count of the number of stars of each type. Our local luminosity function indicates that there are far more faint stars in the Galaxy than bright ones.

<u>Basic Terms Introduced in This Chapter</u>

Spectral classes
HR diagram
Main sequence
"Metal" abundance
Populations I and II
Luminosity function

<u>TOPICS FOR DISCUSSION AND CLASS ACTIVITIES</u>

1. For a long time, astronomy students (who were mostly male) remembered the spectral types of stars by using the mnemonic: "O Be A Fine Girl, Kiss Me Right Now; Smack!" These days, when many women are joining the ranks of astronomers, a new catchphrase seems in order. Owen Gingerich at Harvard has been running a contest in his classes to search for one. Some samples: "O Brutal And Fearsome Gorilla, Kill My Roomate Next Saturday," and "O Bring Another Fully Grown Kangaroo; My Recipe Needs Some." Perhaps your students can come up with even better ones.

2. As mentioned in the "Teaching Notes" above, many students who are not used to graphs and the process of abstraction involved in constructing them will find the HR diagram quite confusing. It helps therefore to approach the subject by means of a graph to which students can relate in a more immediate way.

   For example, you might plot grades on an astronomy exam versus the number of hours spent studying. For most students, there will be a main sequence (which you can get to run the "conventional way" by plotting the abscissa backward). There will be some students (red giants) who will get good grades without a lot of studying by guessing well on multiple-choice questions. Similarly, there will be those (white dwarfs) who study a great deal but have a family quarrel on the morning before the exam and do poorly. Once students understand the concept of the HR diagram, it is a good exercise to ask them to plot one (after giving them a chart with magnitudes and spectral types). (It is often good to include a few stars whose names they might recognize, such as Barnard's Star, Sirius A and B, etc.)

3. Some students may find it clearer for the instructor to discuss the "physical" meanings of the non-main-sequence branches on the HR diagram at this point in the course. That is, that stars in the upper right part of the diagram are cool but bright. How can a star be cool (i.e., each bit of surface emitting relatively small amounts of energy) and yet be bright? There must be a lot of "bits" of surface--i.e., the star must be a giant. Similarly, white dwarfs can be hot but dim by being small.

READING MATERIALS FOR THE STUDENT (FOR CHAPTERS 12 AND 13)

General Books on Stars

Aller, L. *Atoms, Stars and Nebulae* (Harvard Univ. Press, 1971). Somewhat more advanced than the book by Jastrow.

Jastrow, R. *Red Giants and White Dwarfs* (Signet paperback, 1969). Very simplified introduction to stellar and cosmic evolution.

Page, T., and L. Page (eds.). *Starlight--What It Tells About the Stars* (Macmillan, 1967). Articles reprinted from *Sky and Telescope* magazine.

Struve, O., and V. Zebergs. *Astronomy in the 20th Century* (Macmillan, 1962). A historical survey of the development of many of the concepts discussed in this chapter.

See also the books listed for Chapter 18.

Articles

Anderson, J. "The Stars of Very Large Proper Motion," *Sky and Telescope*, August 1969.

Burbidge, M., and G. Burbidge. "Subdwarf Stars," *Scientific American*, June 1961.

Fernie, J. "The Historical Search for Stellar Parallax," *Journal of the Royal Astronomical Society of Canada*, August and October 1975.

Hack, M. "The Hertzsprung-Russell Diagram Today," *Sky and Telescope*, May-June 1966.

Iben, I. "Globular Cluster Stars," *Scientific American*, July 1970.

Joy, A. "M Stars Observed at Mount Wilson," *Astronomical Society of the Pacific Leaflet* #502 (April 1971).

Limber, D. "The Pleiades," *Scientific American*, November 1962.

Mahon, L. "Construction of a Blink Comparator," *Sky and Telescope*, October 1973.

Mihalas, D. "Interpreting Early Type Stellar Spectra," *Sky and Telescope*, August 1973.

Nielsen, A. "Ejnar Hertzsprung--Measurer of Stars," *Sky and*

Telescope, January 1968.
Wilson, O. "A New Scale of Stellar Distances," *Scientific American*, January 1961.

READING MATERIALS FOR THE INSTRUCTOR (FOR CHAPTERS 12 AND 13)

See also the appropriate sections of the general books on stars and their evolution listed for Chapter 18.

Chiu, H., et al. *Stellar Astronomy*, 2 vols. (Gordon and Breach, 1969).
King, I. "Stellar Populations in Galaxies," *Publications of the Astronomical Society of the Pacific*, August 1971.
Morgan, W., and P. Keenan. "Spectral Classification," in *Annual Reviews of Astronomy and Astrophysics*, vol. 11 (1973).
van de Kamp, P. *Principles of Astrometry* (W. H. Freeman and Co., 1967).
van de Kamp, P. "The Nearby Stars," in *Annual Reviews of Astronomy and Astrophysics*, vol. 9 (1971).

AUDIO-VISUAL AIDS (FOR CHAPTERS 12 AND 13)

Films

Invisible Planet (1966, black and white, 30 min). Astronomer Peter van de Kamp is featured in a film about detecting possible unseen companions of stars with large proper motion. (Part of the "Experiment" series; distributed by Indiana University.)
Measuring Large Distances (1959, black and white, 29 min). Triangulation (i.e., parallax) measurements on Earth and in space. (P.S. S.C. film; distributed by Modern Learning Aids.)
The Nature of the Stars (1959, black and white, 29 min). James Pickering discusses stellar classification and the HR diagram in this somewhat elementary film. (Part of the "Astronomy for You" series; distributed by Indiana University.)

Film Loop

Parallax Shift (1969, color, 4 min). (Distributed by Bailey Film Associates.)

SAMPLE EXAM QUESTIONS (MULTIPLE-CHOICE)

1. To construct an HR diagram, astronomers plot stars by their--
   a. apparent magnitude and distance
   b. distance and spectral type
   c. absolute magnitude and spectral type
   d. size and temperature
   e. spectral type and temperature
2. On an HR diagram, most stars lie in a region called the--
   a. red giant branch
   b. supergiant region
   c. white dwarf sequence
   d. subgiant sequence

   e. none of the above
3. An astronomer plots the appropriate characteristics of many stars in a cluster on an HR diagram. He finds that the entire range of stellar types (including bright O stars) is represented on the main sequence. From this he can conclude that the cluster is--
   a. so far away that faint stars in it cannot be observed from Earth
   b. younger than the Sun
   c. older than the Sun
   d. about the same age as the Sun
   e. extremely close to the Sun
4. Population I and Population II stars, as defined in our text, differ mainly in--
   a. distance from the Sun
   b. the presence or absence of a turnoff point on the HR diagrams of clusters
   c. the absolute magnitudes of the dimmer stars
   d. the abundances of the heavier elements
   e. their median income
5. Where would be a good place to look for Population II stars?
   a. in globular clusters
   b. in open clusters
   c. in the immediate neighborhood of the Sun
   d. among the 50 brightest stars in the sky
   e. in the disk of the Galaxy
6. The most abundant type of star in our Galaxy is--
   a. O
   b. A
   c. G (like our Sun)
   d. K
   e. M
7. An instrument commonly used to find stars with large proper motion on photographic plates is called a--
   a. parallaximeter
   b. Schmidt camera
   c. blink comparator
   d. spectrograph
   e. luminosity function

Answers to Multiple-choice Questions

1\. c   2. e   3. b   4. d   5. a   6. e   7. c

SAMPLE EXAM QUESTIONS (ESSAY)

1. Draw an HR diagram, labeling the quantities on the axis and indicating in which direction they are increasing. Draw the various regions in which stars are most often found, and label them. Now draw a second HR diagram, and put in only that one region where *most* stars are found. Indicate where in that region we find (a) the most massive stars, (b) the least massive stars, (c) the brightest stars, (d) the dimmest stars, (e) the hottest stars, and (f) the coldest stars.
2. Why is it that stars with 50 times the mass of our Sun "live" a shorter time than stars with only 1/10 the Sun's mass?
3. You see a newspaper report about some astronomers discovering a

star in our Galaxy with less than 1/10 the mass of the Sun. The article says that this star is done with its nuclear processes and is dying. Why should you be skeptical about such a report?

4. How does the study of cluster HR diagrams show that red giants evolve from main sequence stars?
5. Sirius is a bright A-type star about 2.7 parsecs away from us. Without knowing anything further about its spectrum, which Population do you think it belongs to? Give at least two reasons for your answer.
6. Why is it necessary to do proper-motion studies of faint stars before constructing a luminosity function?

CHAPTER 14
BINARY STARS: TWO ARE BETTER THAN ONE

NOTE BY AUTHOR OF TEXT

Chapters 14 and 15 deal with topics that some astronomy texts set forth in systematic detail just "because they are there." I have insisted instead on subordinating these subjects to the general problems of understanding the Universe. For binaries, the keynote is stated in the third and fourth paragraphs of Chapter 14 and in its final paragraph: we value binaries for the unique sources of information that they offer about stars in general. For eclipsing binaries in particular, the details can be spun out in a fascinating way. On the other hand, this material has less intrinsic interest than the "mainstream" chapters, and much of it can be left out of a course that is pressed for time.

--*I.R.K.*

TEACHING NOTES

In discussing these measurements and calculations (which are sure to seem confusing to many nonscience students), it is good to emphasize how excruciatingly difficult it is to learn about some basic stellar properties in any other way. An observing session focusing on binaries will make the subject a lot more real to the students.

When the details in this chapter are skipped, students probably still ought to read pp. 252-254, if only to learn how common binary systems are. Many students find this a healthy jolt to the Solar System chauvinism with which we are all raised.

SUMMARY OF KEY POINTS

1. About half the stars are *binaries*, with two or more components. If there are more than two, a pairing of components is necessary for stability.
2. Kepler's third law enables us to calculate the masses of double stars from the characteristics of the orbits. Stellar masses are very difficult to determine by other means.
3. Binaries are classified according to the method by which their binary nature can be observed:
   a. *visual binaries* have components that can be seen separately

(thousands are known);

   b. *spectroscopic binaries* can be distinguished by Doppler-shift changes in the spectrum; however this only tells us the line-of-sight velocity and cannot alone give orbital sizes and masses unambiguously;
   c. *eclipsing binaries* are those whose orbits lie edge-on to our line of sight, so that the components have regular variations in their light curve as they eclipse each other. All eclipsing binaries can also be observed as spectroscopic binaries, allowing an unambiguous determination of the stellar masses and radii.

4. There is a strong mass-luminosity correlation for main-sequence stars.
5. Eclipsing-binary light curves can also tell us about--
   a. limb darkening;
   b. tidal distortions, causing a turning of the line of apsides (and this, in turn, can tell us about the distribution of density in the stellar interior).

Basic Terms Introduced in This Chapter

Binary stars
Mass-luminosity relation
Limb-darkening

TOPICS FOR DISCUSSION AND CLASS ACTIVITIES

1. A lecture demonstration about eclipsing binary stars by Neff et al. can be found in *Mercury*, July/August 1972.

2. For classes that have access to small telescopes (or to local amateur astronomers who have telescopes), this may be a good place in the course for observing sessions. "Splitting binaries" has been a favorite occupation of observers for a long time. Most books on observing will have a list of easy binary pairs. The Sky Publishing Corporation (49-50-51 Bay State Road, Cambridge, Mass. 02138) is a good source of books on observational astronomy. A nice list of binary stars for small telescopes can be found in *Sky and Telescope*, Jan.-Feb. 1971.

3. A series of computer-simulated film loops, several of them on stellar motions, is available from Houghton-Mifflin Company (2 Park Street, Boston, Mass. 02107).

4. Binary stars have recently come back into the astronomical spotlight with the discovery that several binaries are X-ray sources. These objects are believed to be pairs that are so close that material can be transferred from one star to another. See "Binary Stars--A Look at Some Interesting Developments," by R. Wilson, *Mercury*, Sept./Oct. 1974, and a number of other articles in "Reading Materials," below.

5. In January 1975, R. Hulse and J. Taylor announced the discovery of a pulsar in a binary system (see, e.g., *Mercury*, May/June 1975, p. 24). In addition to all the other nice things that can be determined from binary objects, this particular pair (with a col-

lapsed object as one component) presents the possibility of checking on some predictions of relativity theory.

READING MATERIALS FOR THE STUDENT

See also the appropriate sections of many of the books listed for Chapters 13 and 18.

Binary Stars in General

Batten, A., and M. Plavec. "Two New Chapters in the Story of U Cephei," *Sky and Telescope*, Sept.-Oct. 1971.

Eggen, O. "Stars in Contact," *Scientific American*, June 1968.

Huang, S. "The Origin of Binary Stars," *Sky and Telescope*, December 1967.

Meeus, J. "Some Bright Visual Binary Stars," *Sky and Telescope*, Jan.-Feb. 1971.

Warner, B. "Six Ultrashort Period Binary Stars," *Sky and Telescope*, December 1972.

Wilson, R. "Binary Stars--A Look at Some Interesting Developments," *Mercury*, Sept./Oct. 1974.

Binary X-ray Sources

Gursky, H., and E. van den Heuvel. "X-ray Emitting Double Stars," *Scientific American*, March 1975.

Jones, C., et al. "X-ray sources and Their Optical Counterparts," *Sky and Telescope*, December 1974.

Thomsen, D. "Where Space Is Sharply Curved," *Science News*, February 23, 1974.

READING MATERIALS FOR THE INSTRUCTOR

Blumenthal, G., and W. Tucker, "Compact X-ray Sources," in *Annual Reviews of Astronomy and Astrophysics*, vol. 12 (1974).

Hjellming, R. "An Astronomical Puzzle Called Cygnus X-3," *Science*, December 14, 1973.

Paczynski, B. "Evolutionary Processes in Close Binary Systems," in *Annual Reviews of Astronomy and Astrophysics*, vol. 9 (1971).

Winkler, L. "Are Most Stars Members of Double Star Systems?" *Physics Today*, September 1967.

AUDIO-VISUAL AIDS

Films or Film Loops

See the films of the Sirius, Xi Ursa Majoris, and Algol systems in the series mentioned in Topic #3, above.

Slides

The slide set on binary and variable stars available from Norton Scientific Company includes several light and velocity curves for binaries.

SAMPLE EXAM QUESTIONS (MULTIPLE-CHOICE)

1. To determine the sum of the masses of a binary star, we need to measure the components'--
   a. temperatures and periods
   b. distance from us and the size of their orbits
   c. periods and the size of their orbits
   d. distance from us and the periastron point
   e. none of the above can help us find the sum of masses.
2. Some binary pairs are so far away from each other that they look like separate field stars. Which of the following would be a way to discover whether such stars are paired?
   a. See if they have the same magnitude.
   b. Determine if they have the same temperature or spectral type.
   c. Use the method of statistical parallaxes.
   d. Measure the size and direction of their proper motions.
   e. Take photographs of them six months apart.
3. Why can observations of a spectroscopic binary not tell us the masses of the components unambiguously?
   a. Spectroscopic binaries have similar spectra, but are not really associated physically.
   b. The lines produced by the dimmer star can never be seen separately.
   c. Doppler shift can tell us only about the component of the orbital motion along the line of sight.
   d. The pairs of lines never come together to enable us to measure their wavelengths unambiguously.
   e. The components always exchange mass, so the distribution of mass between the two stars varies with time.
4. Why don't all binary stars show eclipses?
   a. You need a moon to have eclipses.
   b. You need a planetary system around the stars to have eclipses.
   c. In some cases, one component is simply too large to show eclipses.
   d. Some systems have orbits not aligned edge-on with respect to our line of sight.
   e. Wait just a minute! All binary stars *do* show eclipses.
5. Which of the following can we *not* learn about by observing an eclipsing binary?
   a. the temperatures of the stars
   b. the radii of the stars
   c. limb-darkening
   d. the masses of the stars
   e. tidal distortions on the stars

Answer to Multiple-choice Questions

1. c 2. d 3. c 4. d 5. a

SAMPLE EXAM QUESTIONS (ESSAY)

1. In an episode of a science-fiction TV series, the protagonists enter a quintuple star system with all five components equally spaced. Explain why you would be skeptical about the existence of such a system in "real life." Suggest a more plausible arrangement of the stars.

2. Why is a pair of stars observable as a spectroscopic binary unlikely to be a visual binary? Why would astronomers be happy if more star pairs that are both could be found?
3. Why is it that all eclipsing binaries are spectroscopic binaries, while not all spectroscopic binaries are eclipsing binaries?
4. List four different stellar characteristics that astronomers can find from binary pairs that are both spectroscopic and eclipsing. Briefly describe how each property is determined.

CHAPTER 15

VARIABLE STARS: INCONSTANCY AS A VIRTUE

NOTE BY AUTHOR OF TEXT

As in the preceding chapter, this material is presented for the light that it can shed on the problems of the Universe in general: it shows variable stars as distance indicators and as milestones on the paths of stellar evolution. Even though this approach does lead to a logical division between types of variable stars, no attempt is made to go through all types systematically.

—I.R.K.

TEACHING NOTES

This chapter surveys variable stars from a mainly *observational* perspective. Some instructors may prefer to integrate at least the description of novae and supernovae with the discussion of stellar evolution in Chapter 18. Without the concepts in later chapters, some of the weaker students may find certain topics, such as the evolution of a close binary, somewhat confusing.

Many nonscience students find even simple relations like the period-luminosity law unfamiliar or difficult. For such students, it may be good to give other examples of such relations from more familiar situations. (I like to use functions like the ripening of a piece of fruit with time, or even the level of water in a bathtub.)

SUMMARY OF KEY POINTS

A. *Pulsating variables*. The outer layers of the star oscillate in brightness. The energy to keep the pulsations going is believed to be stored and released in an ionization/recombination zone near the surface of the star.
   1. *Cepheids*—
      a. exhibit a period-luminosity relation that makes them excellent distance indicators (the relation was calibrated using cepheids in the Magellanic Clouds to establish the curve, and then a few cepheids for which distances can be obtained to establish the zero-point);
      b. lie on an "instability strip" in the yellow giant and supergiant region of the HR diagram;

c. are Population I stars.
2. *RR Lyraes*—
    a. vary like cepheids, but are Population II stars;
    b. lie on a region of the horizontal branch of the HR diagram just below the cepheid instability strip;
    c. have a much higher density and therefore a shorter period;
    d. show no period-luminosity relation—they all have nearly the same absolute magnitude;
    e. are consequently good distance indicators to globular clusters and the galactic halo.
3. *Type II cepheids*—
    a. have their own period-luminosity relation, but are Population II stars;
    b. are about 1.5 magnitudes dimmer than "classical" cepheids with the same period.
4. *Miras*—
    a. are long-period variables with light variations of five magnitudes or more;
    b. are M giants or supergiants, and occur in both Populations I and II;
    c. are rare and not well understood.

B. *Explosive variables.*
1. *Novae*—
    a. typically brighten by 12 magnitudes in a few days, with spectra indicating huge expansion velocities for the outer layers;
    b. are blue dwarfs lying between the main sequence and the white dwarf regions of the HR diagram;
    c. are members of close binary systems, in which a main sequence star is transferring material to the dwarf, making its outer regions explosively unstable;
    d. may represent a late stage in the evolution of W Ursae Majoris binaries, whose entire evolution is believed to be altered by mass transfer between the components.
    e. Recurrent novae and dwarf novae have less violent outbursts that recur.
2. *Supernovae*—
    a. are relatively rare explosions of an entire star, releasing energies on the scale of the total previous energy output of the star;
    b. were observed in historical times in 1006, 1054, 1572, and 1604;
    c. are divided into two types. Type I supernovae have a brighter explosion, a slower decline in the light curve, spectra indicating extreme violence, and seem to be associated with old Population I stars. Type II supernovae have an explosion that is less bright, with a faster light-curve decline, and spectra indicating less violence; they seem to be associated with young stars.
3. *Pulsars*—
    a. are very regular, very rapidly varying sources of radio and sometimes visible light and X-ray radiation;
    b. are associated with neutron stars that are formed by supernova explosions.
    c. The best-known example is in the Crab Nebula.

C. *Young variables.*
T Tauri stars—

a. are young G and K stars that vary irregularly;
b. give off tremendous amounts of infrared radiation from a circumstellar dust shell;
c. are probably unstable protostars that have not yet settled onto the main sequence on the HR diagram.

Basic Terms Introduced in This Chapter

Period-luminosity relation
Nova
Limiting surface in a close binary (Roche lobe)

TOPICS FOR DISCUSSION AND CLASS ACTIVITIES

1. As was emphasized in Chapter 12, stellar distances are a key problem for astronomers. This was particularly true around the turn of the century, when many of the modern methods of distance determination were not yet available. Thus the discovery of the period-luminosity law for cepheids was an incredible boon, particularly for those astronomers investigating the structure of the Galaxy. Historically-minded students may find this an interesting topic for a report. See, for example:

   Bok, B. "Harlow Shapley and the Discovery of the Center of our Galaxy," in *The Heritage of Copernicus*, ed. J. Neyman (MIT Press, 1974).
   Struve, O., and V. Zebergs. *Astronomy in the 20th Century* (Macmillan, 1962).

2. The discovery of the period-luminosity law for cepheids was made by Henrietta Leavitt, one of a group of woman research associates working at the Harvard College Observatory around 1900. A discussion of her contribution can lead quite naturally into an examination of the role of women in astronomy and of some of the obstacles they have had to face. For example, it was relatively recently that Margaret Burbidge became the first woman to be permitted to use the 100-inch telescope. (More recently, Dr. Burbidge was appointed the director of the Royal Greenwich Observatory in England, and in 1976 was elected the president of the American Astronomical Society.) She would be an excellent subject for a term paper. For articles about her, see *Smithsonian*, January 1974; *Science Year 1969* (World Book Encyclopedia); and *Mercury*, July/August 1975.

3. Before the 1950's astronomers thought that all cepheids and RR Lyraes could be fit on the *same* period-luminosity diagram. In 1952 Walter Baade reported that his work on M31 with the giant 200-inch telescope showed this assumption to be invalid. As a result, extragalactic distances had to be increased by a factor of two. As some journalists quipped, the size of the Universe doubled overnight! This too would make an interesting report topic, particularly if the student could go back and survey the press reaction at the time. (Begin with pp. 326-330 of *Astronomy in the 20th Century*, by O. Struve and V. Zebergs, Macmillan, 1962.)

4. Historical supernovae in our galaxy are always good subjects for

discussion. Tycho Brahe and Kepler each had the good fortune to see one in the late sixteenth and early seventeenth centuries; Chinese astronomers kept excellent records about the 1054 outburst that produced the Crab Nebula. There has been debate about whether other supernovae have been seen. See, for example:

Stephenson, F. "A Revised Catalog of Pre-telescopic Galactic Novae and Supernovae," *Quarterly Journal of the Royal Astronomical Society*, June 1976.
Stephenson, F., and D. Clark. "Historical Supernovas," *Scientific American*, June 1970.

5. As the enormous energies released by supernova explosions became better understood, a number of people suggested that a nearby outburst might be harmful or deadly for life on Earth. For some recent thinking about this subject, see "Possible Consequences of Nearby Supernova Explosions for Atmospheric Ozone and Terrestrial Life," by M. Ruderman, *Science*, June 7, 1974. As Ruderman points out, our best estimates of supernova rates imply an explosion within 50 light years of us each few hundred million years or less. Thus a significant number of such events should have occurred during the history of the Earth. Students may enjoy researching some of the suggestions about the historical effects of supernovae, which range from accelerating the pace of mutations to killing off the dinosaurs.

6. The text merely introduces the subject of pulsars. Students who want to learn more about these fascinating and controversial objects can be referred to the "Reading Materials" listed below. In particular, students may want to look at the gradual slowing-down and occasional speeding-up of the fastest pulsar frequencies, and what these observations can tell us about pulsar structure and energetics.

7. With reference to Topic #2, above, we should note that pulsars were originally found in rather confusing data by a woman graduate student, Jocelyn Bell. Her story is chronicled by N. Wade in "Discovery of Pulsars: A Graduate Student's Story," *Science*, August 1, 1975.

8. In 1975, R. Hulse and J. Taylor at the University of Massachusetts announced the discovery of a pulsar in a binary system, whose other member may also be a collapsed object. Investigation of this fascinating object is continuing. (The discovery was announced in *Astrophysical Journal Letters*, vol. 195, p. L51.)

READING MATERIALS FOR THE STUDENT

Many of the books cited for Chapters 13 and 18 contain sections dealing with the topics of this chapter.

Pulsating Stars, Novae, etc.

Bok, B., and P. Bok., (eds.). *The Milky Way*, 4th ed. (Harvard Univ. Press, 1974). [See Chapter 5.]
Glasby, J. *Variable Stars* (Harvard Univ. Press, 1968).

Kraft, R. "Pulsating Stars and Cosmic Distances," *Scientific American,* July 1959.
Kraft, R. "Exploding Stars," *Scientific American,* April 1962.
Kraft, R. "Are All Novae Binary Stars?" *Astronomical Society of the Pacific Leaflet* #418 (April 1964).
Lovell, B. "Radio Emitting Flare Stars," *Scientific American,* August 1964.
"More About Nova Cygni," *Sky and Telescope,* November 1975.
Percy, J. "Pulsating Stars," *Scientific American,* June 1975.
Struve, O. "The Pulsating Star RR Lyrae," *Sky and Telescope,* June 1962.
Thomsen, D. "A Funny Thing Happened on the Way to the Main Sequence," *Science News,* January 10, 1976. [About T Tauri stars.]
Warner, B., and R. Nather. "High Speed Photometry of Cataclysmic Variables," *Sky and Telescope,* February 1972.

Supernovae and Their Remnants

Bok, B. "The Gum Nebula," *Sky and Telescope,* August 1971.
Brandt, J., and T. Stecher. "An Invisible Cloud," *Smithsonian,* February 1972. [About the Gum Nebula.]
Burbidge, G. "Dissecting the Crab," *Natural History,* October 1970.
Charles, P., and J. Culhane. "X-rays from Supernova Remnants," *Scientific American,* December 1975.
Gorenstein, P., and W. Tucker. "Supernova Remnants," *Scientific American,* July 1971.
Maran, S. "The Gum Nebula," *Scientific American,* December 1971.
Miller, E., and J. Muzzio. "Mosaics of a Southern Nebula," *Sky and Telescope,* February 1975. [About the Gum Nebula.]
Schramm, D., and W. Arnett. "Supernovae: The Origin of the Chemical Elements, Cosmic Rays, Neutron Stars and Maybe Even Black Holes," *Mercury,* May/June 1975.

Pulsars

Green, L. "Pulsars Today," *Sky and Telescope,* Nov.-Dec. 1970.
Green, L. "Starquakes: Have They Been Observed?" *Sky and Telescope,* February 1971.
Gursky, H., and E. van den Heuvel. "X-ray Emitting Double Stars," *Scientific American,* March 1975.
Hewish, A. "Pulsars," *Scientific American,* October 1968.
Ostriker, J. "The Nature of Pulsars," *Scientific American,* January 1971.
Pacini, F., and M. Rees. "Rotation in High Energy Astrophysics," *Scientific American,* February 1973.
Trimble, V. "The Crab Nebula and Pulsar," in *The Emerging Universe,* ed. W. Saslaw and K. Jacobs (Univ. Press of Virginia, 1972).
Wade, N. "Discovery of Pulsars: A Graduate Student's Story," *Science,* August 1, 1975.

READING MATERIALS FOR THE INSTRUCTOR

Blumenthal, G., and W. Tucker. "Compact X-ray Sources," in *Annual Reviews of Astronomy and Astrophysics,* vol. 12 (1974).
DiLavore, P., and J. Wayland. "Pulsars for the Beginner," *Physics Teacher,* May 1971.
"Discovery of the First Optical Pulsar," an entire section in *Adventures in Experimental Physics,* Alpha vol. (January 1972).

Ginzburg, V., and V. Zheleznyakov. "On the Pulsar Emission Mechanisms," in *Annual Reviews of Astronomy and Astrophysics*, vol. 13 (1975).
Hewish, A. "Pulsars and High Density Physics" [Nobel-Prize lecture], *Science*, June 13, 1975.
Hjellming, R. "An Astronomical Puzzle Called Cygnus X-3," *Science*, December 14, 1973.
Kukarkin, B., (ed.). *Pulsating Stars* (Wiley, 1975).
Maran, S., et al. "The Gum Nebula: A New Kind of Astronomical Object," *Physics Today*, September 1971.
National Aeronautics and Space Administration. *The Gum Nebula and Related Problems* (N.A.S.A. Publication SP-332, 1972). [Proceedings of a 1971 conference. Available from National Technical Information Service, Springfield, Va. 22151.]
Oke, J., and L. Searle. "The Spectra of Supernovae," in *Annual Reviews of Astronomy and Astrophysics*, vol. 12 (1974).
Paczynski, B. "Evolutionary Processes in Close Binary Systems," in *Annual Reviews of Astronomy and Astrophysics*, vol. 9 (1971).
Ruderman, M. "Pulsars: Structure and Dynamics," in *Annual Reviews of Astronomy and Astrophysics*, vol. 10 (1972).
Shklovskii, I. *Supernovae* (Wiley Interscience, 1968).
Strom, S., et al. "Young Stellar Objects and Dark Interstellar Clouds," in *Annual Reviews of Astronomy and Astrophysics*, vol. 13 (1975).
Tinsley, B. "What Stars Become Supernovae?" *Publications of the Astronomical Society of the Pacific*, December 1975.
Woltjer, L. "Supernova Remnants," in *Annual Reviews of Astronomy and Astrophysics*, vol. 10 (1972).

AUDIO-VISUAL AIDS

Films

The Crab Nebula (1970s, color, 52 min). This superb BBC-TV film treats the history of our understanding of this supernova remnant as a most exciting detective story. Fine photography, well-edited interviews, and effective models all work together to create one of the best astronomy films ever made. (Distributed by Time-Life Films.)

Exploring the Milky Way (1967, color, 29 min). George Preston narrates and is featured in this A.A.S. film of a typical night at the Lick Observatory. Includes a slightly rushed discussion of cepheid and Mira variables and their use in charting the structure of our Galaxy. (Distributed by Modern Learning Aids.)

Milky Way (1953, black and white, 10 min). Includes a discussion of Nova Aquilae from 1918 to 1931. (Produced by the International Screen Organization.)

Slides

Color and black-and-white slides of some of the objects discussed in this chapter are available from the Hale or Lick Observatories and from some secondary sources. For example, Hale lists the Crab Nebula (S-22) and a supernova in NGC 5236 (SP-55).

Norton Scientific Company has a whole set of slides about binary and variable stars, including light curves for several well-known variables and the cepheid period-luminosity law.

SAMPLE EXAM QUESTIONS (MULTIPLE-CHOICE)

1. Cepheids are good distance indicators because there is a relationship between their period and their—
   a. distance
   b. apparent magnitude
   c. absolute magnitude
   d. central temperature
   e. location in the Galaxy
   f. ZIP codes
2. To discover the period-luminosity relation, Henrietta Leavitt had to observe a large group of cepheids at about the same distance from us. She found them in—
   a. the Andromeda galaxy
   b. a Magellanic Cloud
   c. the Crab Nebula
   d. an open cluster
   e. a globular cluster
   f. the Yellow Pages

[NOTE: Questions 1 and 2 probably should not be used on the same exam.]

3. Astronomers believe that the energy for keeping the pulsations going in a cepheid is stored and released in—
   a. the instability strip
   b. a period-luminosity sequence
   c. the core of the star, which is heated and cooled
   d. an ionization and recombination zone
   e. a hot atmosphere right above the surface of the star
4. RR Lyrae stars differ from cepheids in that they—
   a. are Population II stars
   b. do not pulsate
   c. have shorter periods
   d. cannot be placed on the HR diagram
   e. more than one of the above
5. The fact that RR Lyrae stars lie on a horizontal branch on the HR diagram indicates that they have pretty much the same—
   a. luminosity
   b. spectral type
   c. period
   d. distance
   e. apparent magnitude
   f. average on astronomy tests
6. Long-period variable stars are also called—
   a. cepheids
   b. novae
   c. Type II cepheids
   d. Miras
   e. RR Lyrae
7. All novae are—
   a. red stars
   b. cepheids
   c. close binaries
   d. sites for the formation of pulsars
   e. outside our Galaxy
8. Today astronomers believe that the progenitors of novae are—
   a. W Ursa Majoris stars
   b. U Geminorum stars

c. RR Lyrae stars
d. Type II cepheids
e. supernovae

9. You are a young and impatient graduate student in astronomy searching for a new supernova. Where would be the best place to search?
   a. in the Crab Nebula
   b. among the 50 nearest stars
   c. on the far side of our Galaxy
   d. where pulsars are found
   e. in some other, large galaxy
10. Suppose that an extremely bright supernova went off in our Galaxy 20,000 light years away and 20,000 years ago. The most likely reason that we would not be able to see its light right now is that—
   a. supernovae are brilliant for less than a day, and chances are we'd miss it
   b. we cannot see anything at all in the Universe as far away as 20,000 light years
   c. no supernova would grow bright enough to be seen at that distance
   d. dust between us would obscure the light
   e. supernovae only occur in other galaxies
11. At the center of the Crab Nebula, astronomers have found a regularly variable star. This turns out to be a—
   a. cepheid
   b. pulsar
   c. T Tauri star
   d. RR Lyrae star
   e. Mira
12. Which of the following variable-star types generally would be on the main sequence of the HR diagram?
   a. T Tauri stars
   b. cepheids
   c. novae
   d. RR Lyrae stars
   e. None of the above lie on the main sequence.

Answers to Multiple-choice Questions

1. c   2. b   3. d   4. e   5. a   6. d   7. c   8. a   9. e   10. d
11. b   12. e

SAMPLE EXAM QUESTIONS (ESSAY)

1. You are an astronomer with a good telescope and auxiliary equipment at your disposal. Describe briefly how you would search for a cepheid variable, and what steps you would take to find its distance.
2. Discuss the various measurements that had to be made to calibrate the period-luminosity relation for cepheids. Why was it worth all that trouble?
3. An amateur astronomer down the block discovers a variable star and asks you for help in identifying which type of variable it is. What observations would you advise him to make, and Why?
4. How do RR Lyrae and Type II cepheids differ from "classical"

cepheids?

5. You are a guest astronomer at a major observatory. Suddenly you receive a phone call announcing that a nova has just "gone off" in an area of the sky that you are particularly well situated to see. Describe three observations that you would make during the nights that follow.
6. Briefly describe how a close W Ursa Majoris system might evolve into a nova.
7. Distinguish between Type I and Type II supernovae in three different ways.
8. Why do astronomers believe that pulsars are the result of supernova explosions? [NOTE: This question can be deferred until Chapter 18.]
9. Describe and explain two observations that indicate that T Tauri variables are protostars that have recently formed.

CHAPTER 16
THE SECRETS OF STARLIGHT: UNRAVELING THE SPECTRUM

NOTE BY AUTHOR OF TEXT

This is a key chapter but a difficult one. The most important idea is that much of the way in which the world works is due to the atoms that make it up. Stellar spectra are used here as a laboratory for the study of atomic behavior. As elsewhere in the book, the thrust is along two parallel lines: understanding how the Universe works, and seeing how we manage to understand the stars. In the final section, note the breakpoint near the end of page 307. The first two pages introduce the uncertainty principle, to which every student should be exposed. The remainder of the section, however, discusses details that can be omitted from a short course.

--I.R.K.

TEACHING NOTES

A great deal of material that some students will find difficult is summarized in this chapter, often quite concisely. For classes with little science and math background, these areas must be covered very slowly, with many examples, exercises, and demonstrations. (For the latter, see Topic #1, below.)

Students will probably have most trouble with the section on radiation laws, particularly when they must keep several variables in mind simultaneously. For example, the terse paragraph on calculating stellar diameters on p. 296 would require a great deal of patient elucidation to be clear to most of my nonscience students.

Similarly, the discussion on the uncertainty principle will not mean much to students who have not thought much before about such problems and issues in physics. Some of the films on quantum mechanics listed for Topic #3, below, may help, but unless the instructor has time to discuss the uncertainty principle at some length, the section on line profiles is perhaps best skipped altogether for nonscience majors.

By the way, although spectra are clearly defined only in this chapter, they have been discussed before. Students may wish to review pp. 21-22 on the spectrograph, pp. 228-229 on the Doppler effect, and pp. 235-236 on spectral types before tackling this chapter.

SUMMARY OF KEY POINTS

Continuous Spectrum

1. For a black body, the distribution of radiation intensity with wavelength depends only on the temperature and is specified by Planck's law.
2. Wien's law states that the wavelength at which radiation is most intense is inversely proportional to the absolute temperature; this law can be used to measure temperature of distant objects like stars.
3. Because a star's color is determined by the distribution of its light with wavelength, star colors are a good rough index of star temperatures.
4. Stefan's law states that the total rate of radiation from a body is proportional to its surface area times the absolute temperature to the fourth power. It can be used to calculate a star's diameter from its luminosity and temperature.
5. The continuous spectrum is produced by bound-free and free-free transitions between atoms and free electrons.

Line Spectrum

1. Electrons in an atom can absorb and emit energy in discrete packets, called photons. The energy of a photon is inversely proportional to the wavelength. The intensity of a beam of light is determined by the number of photons per second.
2. The spacings of the energy levels for the electrons determine the energies (and thus the wavelengths) of the photons a given atom can absorb and emit. Each atom thus has a characteristic pattern of lines in the spectrum that can be used to identify the element.
3. In stars, absorption lines are formed in the continuous spectrum when radiation from the hotter interior passes through the cooler atmosphere of the star. Which of the possible lines of each element are present depends on the temperature of the star.
4. Because metals hold their electrons more loosely than nonmetals, at visible wavelengths we see mostly lines from nonmetallic elements in hot stars and from metals in cool stars.
5. In photoionization, an electron receives so much energy that it can break away from the atom. Some such electrons recombine, and that atom then emits a photon whose energy may have a range of values. In free-free transitions, the atom interacts with a passing electron, absorbing or emitting a photon at the expense of the electron's energy of motion.

Line Profiles

A spectral line is spread out over a range of wavelengths for several reasons.

1. It has a natural width due to the "spreading" of the energy levels by the quantum-mechanical uncertainty principle.
2. Collisions can shorten the time an electron spends in an energy level and thus broaden the energy spread of the level and the line.
3. The Stark effect will broaden lines if there are significant numbers of ions perturbing nearby atoms.
4. Motions of the atoms (either thermal motion or large-scale turbulent motions) can broaden line profiles because of the Doppler effect.

5. The rotation of the star also can broaden the line profile in a measurable way; this is the only way we can measure stellar rotation.

Basic Terms Introduced in This Chapter

Spectrogram
Continuous spectrum
Black body
Planck's law
Wien's law
Stefan's law
Color index
Ion
Photon
Bound-free transition
Free-free transition
Uncertainty principle
Natural width (of a spectral line)
Collisional broadening
Stark effect
Doppler broadening

TOPICS FOR DISCUSSION AND CLASS ACTIVITIES

1. Laboratory or classroom demonstrations about the physical principles discussed in this chapter can be extremely useful at this point. For example:
    a. Repeating the classical double-slit interference experiment *and* the photoelectric-effect experiment in sequence can be a dramatic way of stressing the dual nature of light.
    b. Students can be provided with inexpensive transmission gratings to view spectra for themselves. Edmund Scientific Company markets inexpensive sets of grating slides that can be used in conjunction with a thin light source. Slightly more expensive are hand-held cardboard spectroscopes usable with any sort of light source.

    Some basic but important considerations about teaching stellar spectra are emphasized in "Effective Astronomy Teaching: Intellectual Development and Its Implications," by D. Schatz and A. Lawson, *Mercury*, July/August 1976.

2. Two lab exercises suitable for class use and relevant to this chapter are available from Sky Publishing Corporation.

    Spectral Classification (LE-2) lets the student compare 30 "unknown" spectra with standard ones.
    The Rotation of Saturn and Its Rings (LE-3) allows the student to measure the rotation of both planet and ring from characteristics of the spectral lines.

3. There are many books for students interested in learning more about quantum mechanics and the uncertainty principle. One that I have found particularly effective for nonscience majors is A. Baker's *Modern Physics and Antiphysics* (Addison-Wesley paperback, 1970). Several films are useful for clarifying some of the

rather abstract and bizarre notions of quantum mechanics, including the following:

A New Reality (1965, color, 51 min). Made in honor of Niels Bohr, this is a good introduction to concepts of atomic structure. (Distributed by the International Film Bureau.)
Probability and Uncertainty (1964, black and white, 56 min). In this sequence from the "Character of Physical Law" lectures delivered by Richard Feynman in 1964, the Nobel laureate discusses some basic experiments that lead to the uncertainty principle. (A BBC film; distributed by Educational Development Center.)

4. On p. 295, the text discusses the fact that our eyes "were designed to use sunlight." Because a great deal of the process of evolution involves adaptation to a particular environment, it is interesting to speculate how life may have adapted to environments that are significantly different from the Earth. See, for example, some intriguing speculation by B. Dalzell in "Exotic Bestiary for Vicarious Space Voyagers," *Smithsonian*, October 1974. (See also Topic #2 for Chapter 2.)

5. Although there have been a few false alarms, there is at present no evidence from the study of stellar and galactic spectra for the existence of any lines not ascribable to elements and compounds we know from our understanding of chemistry here in our Solar System. The fact that all the thousands of spectral lines from local objects are also found in the most distant sources of radiation is a nice and powerful argument for the uniformity of the laws of nature, as discussed in Chapter 1. (See Topic #7 for Chapter 1.)

6. The text mentions that the detailed analysis of line profiles is really our only way of knowing about stellar rotation. It might be useful when discussing this to remind students of the importance of stellar rotation in modern theories of planet formation, as discussed in Chapter 11.

READING MATERIALS FOR THE STUDENT

Aller, L. *Atoms, Stars and Nebulae*, rev. ed. (Harvard Univ. Press, 1971). [The first few chapters summarize spectra and their applications.]
Buscombe, W. "What Is Peculiar About A Stars?" *Astronomical Society of the Pacific Leaflet* #508 (October 1971).
Hack, M. "Stellar Rotation and Atmospheric Motion," *Sky and Telescope*, August, September, and October 1970.
Hewish, A., (ed.). *Seeing Beyond the Visible* (American Elsevier, 1970). [The first three chapters are a general introduction to radiation and atoms.]
Menzel, D. "The History of Astronomical Spectroscopy," in "International Conference on Education in and the History of Modern Astronomy," ed. R. Berendzen, *Annals of the New York Academy of Science*, August 25, 1972.
van Heel, A., and C. Velzel. *What Is Light?* (McGraw-Hill paperback,

1968). [An almost completely nontechnical, nonmathematical, well-illustrated introduction to light, waves, and photons.]

Whitney, C. "New Trends in Spectral Classification," *Sky and Telescope*, June 1968.

In addition, almost any introductory astronomy or physics textbook devotes some attention to spectra, atoms, and radiations. Students should be urged to make use of local libraries to find further explanations of topics introducted in this chapter.

READING MATERIALS FOR THE INSTRUCTOR

Bertojo, M. "Isotopic Abundances and Their Variations Within the Galaxy," *Science*, May 10, 1974.

Hummer, D., and G. Rybicki. "The Formation of Spectral Lines," in *Annual Reviews of Astronomy and Astrophysics*, vol. 9 (1971).

Jefferies, J. *Spectral Line Formation* (Blaisdell, 1968).

Keenan, P. "Classification of Stellar Spectra," in *Basic Astronomical Data*, ed. K. Strand (Univ. of Chicago Press, 1963).

Mihalas, D. *Stellar Atmospheres* (W. H. Freeman and Co., 1970). [A highly regarded graduate-level textbook.]

Preston, G. "The Chemically Peculiar Stars of the Upper Main Sequence," in *Annual Reviews of Astronomy and Astrophysics*, vol. 12 (1974).

AUDIO-VISUAL AIDS

Films

Dark Lines to the Planets (1964, color, 19 min). Although this film is mainly about determining the compositions of planetary atmospheres, it does discuss absorption spectroscopy in some detail. (Pennsylvania State University film.)

Stars (1971, color, 16 min). Starts with a brief history of stellar astronomy, and then discusses measuring the composition and temperatures of stars from their light. (Produced by Haberstroh Studios; distributed by McGraw-Hill Films.)

What Are Stars Made Of? (1960, color, 16 min). Armin Deutsch discusses the work of astronomers in general, and taking and interpreting spectra in particular. (Distributed by International Film Bureau.)

Film Loops

Absorption Spectra (1969, color, 4 min). Shows the spectra of sodium and several biological molecules. (Part of the "Demonstrations in Physics" series available from Franklin Miller at Kenyon College, Gambier, Ohio.)

The Spectroscopy Series (1972, color). Includes loops on the nature of spectra, emission spectroscopy, visible ultraviolet absorption spectroscopy, atomic absorption, infrared spectroscopy, etc. (Produced by J. Davis at the University of South Florida; distributed by BFA Educational Materials.)

Spectrum of the Hydrogen Atom (1967, color, 3 min). Shows the formation of a line spectrum. (Produced by Halas and Batchelor;

distributed by Encyclopaedia Britannica Educational Films.)

Slides

Slides of stellar spectra are available from the Hale, Lick, and Yerkes observatories. Norton Scientific Company markets a set of slides that includes not only spectra, but also the radiation laws.

Audio Tape

The Development of the Uncertainty Principle (17 min). Werner Heisenberg discusses the origin of the principle for which he was responsible. (Distributed by Spring Green Multimedia.)

SAMPLE EXAM QUESTIONS (MULTIPLE-CHOICE)

1. For a black body, the wavelength distribution of emitted radiation depends only on the object's--
   a. mass
   b. visible luminosity
   c. temperature
   d. chemical composition
   e. electrical charge
2. For a given temperature, Planck's law specifies the distribution of radiation intensity in a black body as a function of--
   a. wavelength
   b. stellar mass
   c. depth within the star
   d. distance from the object
   e. the star's position on the HR diagram
3. The statement that the predominant wavelength of the radiation from a black body is inversely proportional to the absolute temperature is called--
   a. Stefan's law
   b. Wien's law
   c. King's law
   d. Doppler's law
   e. the Balmer law
4. The walls of the room in which you are suffering through this exam radiate predominantly in which region of the electromagnetic spectrum?
   a. radio
   b. infrared
   c. visible light
   d. ultraviolet
   e. X-rays
5. Our Sun emits its most intense radiation in which region of the electromagnetic spectrum?
   a. radio
   b. infrared
   c. visible light
   d. ultraviolet
   e. X-rays
6. A photon is--
   a. a positively charged nuclear particle
   b. an instrument that astronomers attach to a telescope to help record light

c. an element with a high atomic number
d. a packet of radiation energy
e. a new kind of Japanese motorcycle

7. All photons moving through space have the same--
a. wavelength
b. frequency
c. energy
d. color
e. speed

8. Which of the following types of photons has the highest energy?
a. X-rays
b. visible light
c. infrared
d. gamma rays
e. ultraviolet

9. In a beam of light of the same color, each photon has the same--
a. wavelength
b. energy
c. speed
d. frequency
e. all of the above

10. To determine what the outer regions of a star are made of, astronomers would measure the star's--
a. continuous spectrum
b. spectral absorption lines
c. predominant wavelength
d. luminosity
e. surface temperature

11. A photon is emitted in a free-free interaction. Where did the energy for that photon come from?
a. The electron went from a higher to a lower energy level in its atom.
b. The electron went from a lower to a higher energy level in its atom.
c. The free electron lost some energy of motion.
d. The free electron recombined with the ion.
e. An electron was photoionized.

12. In the atmosphere of a given star, whole layers of gas are in large-scale motion. The resulting broadening of the spectral lines coming from this region will be caused by--
a. the Stark effect
b. the Doppler effect
c. Wien's law
d. Stefan's law
e. the Einstein effect

13. To measure the rotation speed of a star, astronomers examine--
a. its variation in luminosity as it turns
b. its temperature
c. the number of hydrogen lines observable in its spectrum
d. the line profiles in its spectrum
e. its position on the HR diagram

<u>Answers to Multiple-choice Questions</u>

1. c 2. a 3. b 4. b 5. c 6. d 7. e 8. d 9. e 10. b
11. c 12. b 13. d

SAMPLE EXAM QUESTIONS (ESSAY)

1. To a reasonable approximation, a star like Betelgeuse can be considered a black body. How can this be, since Betelgeuse is a red star?
2. Using sensitive equipment and a new telescope in the southern hemisphere, an astronomer investigates a star no one has really examined before. Describe what observations and data manipulations he will perform to measure the star's temperature.
3. List and briefly explain the steps required to calculate the diameter of a star from the radiation it emits.
4. Give a guided tour of the atom designed for someone who has not had the benefit of this course. Be sure to include (a) where most of the mass is; (b) the electric charges of the various parts; and (c) how the parts remain together.
5. I have two beams of light. Beam 1 has 1,000,000 photons per second at a wavelength of 5,000 Å, whereas beam 2 has 3,000,000 photons per second at a wavelength of 4,000 Å. (a) Which beam has greater intensity? (b) Which beam has photons of greater energy? (c) Which beam travels faster? Explain your answers!
6. Astronomers claim that they have not discovered any elements in the stars that are not known in our Solar System. Explain how astronomers reach this conclusion.
7. Why are *all* the possible spectral lines of an element that is found in a given star not present in the star's spectrum? What determines which lines are present?
8. List and discuss three different mechanisms that can broaden spectral lines in stars. What characteristics of stars can we learn about from the way this broadening occurs?

CHAPTER 17
THE SUN: OUR OWN STAR

NOTE BY AUTHOR OF TEXT

This largely descriptive chapter about the Sun is a light interlude between the heavy chapters that precede and follow it. It can be omitted from a short course. Although details of solar phenomena can be elaborated ad infinitum, I have tried instead to keep a stellar point of view, asking what the Sun can tell us about stars in general.

—I.R.K.

SUMMARY OF KEY POINTS

Solar Surface

1. The photosphere is the region where the Sun becomes opaque and from which the radiation comes.
2. Because at oblique angles we see higher and cooler gas, the limb of the Sun appears darker than the center of the disk.
3. The surface shows dark and light granulation as hotter material rises and cooler material sinks because of ionization in a hydrogen convection zone just beneath the surface.
4. Sunspots are larger areas that look darker because the material on them is cooler. Sunspot phenomena are connected with the solar magnetic field. The drifting of sunspots shows the rotation of the Sun.

Chromosphere and Corona

1. The chromosphere is a narrow region at the base of the corona where hydrogen and helium are excited but not completely ionized. It is visible during an eclipse.
2. The corona is the region of hot thin gas that surrounds the Sun. Here hydrogen and helium are completely ionized.
3. Prominences are arcs of material, extending from the chromosphere into the corona along magnetic field lines, in which the hydrogen and helium are cool enough not to be ionized.
4. The solar wind is the flow of gas away from the Sun, probably powered by shock waves generated by the motions of the granulation in the photosphere.

Solar Activity

1. Flares are sudden releases of energy in very small areas of the Sun.
2. Sunspots, prominences, and flares tend to concentrate in active regions whose number and strength varies on a cycle of about 11 years. This cycle is probably connected with the effect of the Sun's differential rotation on its magnetic field.
3. Strong solar activity can increase the ionization of the Earth's ionosphere and thus disrupt radio communications.

Solar Radio Astronomy

1. The Sun becomes opaque to radio waves in the corona. Thus the radio Sun looks bigger than the optical Sun.
2. When solar activity is high, radio "bursts" (which can be as much as a million times stronger than the "quiet Sun") are frequently observed.

Basic Terms Introduced in This Chapter

Photosphere
Limb-darkening
Granulation
Sunspots
Corona
Chromosphere
Prominences
Flares

TOPICS FOR DISCUSSION AND CLASS ACTIVITIES

1. It is useful to give students an idea of the relative scale of the Sun and the Earth. For example, you might point out that the Earth could fit comfortably into many of the prominences.

2. On p. 323, mention is made of government support for solar astronomy, stemming from a fear of disrupted radio communications. An interesting but rather difficult student project might be to look into the recent history of and politics behind the funding of solar research.

3. If the time is right and equipment is available, one basic but pretty effective class project would be to observe sunspots, either with a filter or by carefully projecting an image of the solar surface.

4. The 11-year sunspot cycle has intrigued investigators ever since it was first discovered in the mid-1800s. It clearly seems correlated with auroras and magnetic storms on Earth. Other suggestions include a connection between the cycle and such disparate phenomena as Earth climate, the quality of good French wines, business cycles, etc. Many such claims are unsubstantiated and would make good subjects for an investigative report.

5. There is some evidence that sunspot activity was significantly reduced from about 1645 to 1715, with some effects on the Earth's climate. See "The Maunder Minimum," by J. Eddy, *Science*, June 18, 1976.

6. A lot of interest has focused (if you'll pardon the expression) on using solar power for our planet's energy needs. This is usually a popular paper topic. The following references could serve as starting places.

Douglas, J. "Breaking the Solar Energy Logjam," *Science News*, April 13, 1974.
Meinel, A., and M. Meinel. "Physics Looks at Solar Energy," *Physics Today*, February 1972.
Meinel, A., and M. Meinel. *Applied Solar Energy* (Addison-Wesley, 1976). [A somewhat technical textbook on the topic.]
O'Neill, G. "The High Frontier," *CoEvolution Quarterly*, Fall 1975. [A popular summary of a proposal (apparently being considered seriously by various government agencies) for collecting solar energy at huge orbiting space habitats and beaming it down to Earth.]
"An Orbiting Solar Power Station," *Sky and Telescope*, April 1975.

READING MATERIALS FOR THE STUDENT

Altschuler, M. "The Geometry of the Coronal Magnetic Field," *Sky and Telescope*, March 1971.
Ashbrook, J. "The Sun's Shape," *Sky and Telescope*, August 1975.
Athay, G. "The Solar Chromosphere," *Scientific American*, February 1962.
Babcock, H. "The Magnetism of the Sun," *Scientific American*, February 1960.
Brandt, J. "The Solar Wind," *Smithsonian*, January 1973.
Bray, R., and R. Loughead. *Sunspots* (Barnes and Noble, 1965).
Driscoll, E. "Peeking Beneath the Sun's Skirts," *Science News*, August 18-25, 1973. [About Skylab results.]
Gibson, E. "The Sun as Never Seen Before," *National Geographic*, October 1974. [About Skylab results.]
Gingerich, O. "The Sun," in *Man and Cosmos*, ed. J. Cornell and E. Hayes (Norton, 1975).
Howard, R. "The Rotation of the Sun," *Scientific American*, April 1975.
Livingston, W. "Magnetic Fields on the Quiet Sun," *Scientific American*, November 1966.
Livingston, W. "Determining the Rotation of the Sun," *Astronomical Society of the Pacific Leaflet* #484 (October 1969).
Livingston, W. "Measuring Solar Photospheric Magnetic Fields," *Sky and Telescope*, June 1972.
Maran, S. "Solar Bubbles," *Natural History*, April 1975.
Maran, S., and R. Thomas. "The OSO-7 Year of Discovery," *Sky and Telescope*, January 1973.
Maxwell, A. "Radio Emissions from Solar Flares," *Sky and Telescope*, July 1973.
Menzel, D., and J. Pasachoff. "Solar Eclipse: Nature's Super Spectacular," *National Geographic*, August 1970.
Parker, E. "The Solar Wind," *Scientific American*, April 1964.
Parker, E. "The Sun," *Scientific American*, September 1975.
Pasachoff, J. "The Solar Corona," *Scientific American*, October 1973.
"Recent Solar Activity," *Sky and Telescope*, September 1974.
"Some Results from Skylab's Solar Experiments," *Sky and Telescope*, July 1974.
Starr, V., and P. Gilman. "The Circulation of the Sun's Atmosphere,"

Scientific American, January 1968.
Sykes, G. "A Sonoran Adventure," *Mercury*, Jan./Feb. 1972. [About a 1923 eclipse-observing expedition.]
Tandberg-Hanssen, E. "Solar Prominences and Their Magnetic Fields," *Sky and Telescope*, August-Sept. 1971.
Thomas, R., and S. Maran. "What Two Sun-observing Satellites Tell Us," *Sky and Telescope*, May 1971. [About OSO-5 and OSO-6.]

READING MATERIALS FOR THE INSTRUCTOR

Beckers, J. "Solar Spicules," in *Annual Reviews of Astronomy and Astrophysics*, vol. 10 (1972).
Bradford, H. "Plasma-frequency Radiation in Astrophysics: What We Learn from Solar Radio Observations," *Journal of the Royal Astronomical Society of Canada*, August 1975.
Brandt, J. *Introduction to the Solar Wind* (W. H. Freeman and Co., 1970).
Culhane, J., and L. Acton. "The Solar X-ray Spectrum," in *Annual Reviews of Astronomy and Astrophysics*, vol. 12 (1974).
Gibson, E. *The Quiet Sun* (N.A.S.A. Publication SP-303, 1973).
Gilman, P. "Solar Rotation," in *Annual Reviews of Astronomy and Astrophysics*, vol. 12 (1974).
Howard, R. "Recent Solar Research," *Science*, September 29, 1972.
Meadows, A. *Early Solar Physics* (Pergamon, 1970). [Reprints major papers in the development of solar physics, together with a long introduction.]
Noyes, R. "Ultra-violet Studies of the Solar Atmosphere," in *Annual Reviews of Astronomy and Astrophysics*, vol. 9 (1971).
Noyes, R. "New Developments in Solar Research," in *Frontiers of Astrophysics*, ed. E. Avrett (Harvard Univ. Press, 1976).
Stein, R., and J. Leibocher. "Waves in the Solar Atmosphere," in *Annual Reviews of Astronomy and Astrophysics*, vol. 12 (1974).
Svestka, Z. "Spectra of Solar Flares," in *Annual Reviews of Astronomy and Astrophysics*, vol. 10 (1972).
Wild, J., and F. Smerd. "Radio Bursts from the Solar Corona," in *Annual Reviews of Astronomy and Astrophysics*, vol. 10 (1972).
Zirin, H. *The Solar Atmosphere* (Blaisdell, 1966).

AUDIO-VISUAL AIDS

Films

The Active Sun (1969, color, 27 min). Comprehensive film about solar activity and how we learn about it. (Part of the "Sun and the Earth" series, produced by the National Academy of Science; distributed by McGraw-Hill Films.)
The Living Sun (1967, black and white, 30 min). National Educational Television documentary about equipping and flying an airplane to observe a solar eclipse. (Distributed by Indiana University.)
Mystery of the Sun (1960, black and white, 25 min). Emphasizes the use of rockets in solar research. Note the date. (A CBS film.)
Partnership into Space: Mission HELIOS (1975, color, 27 min). A documentary about the development and launch of the joint U.S.-

German satellite to study the Sun. (N.A.S.A. film HQ-254).

The Quiet Sun (1968, color, 29 min). General information about the Sun, its structure, and its energy source. (Part of the "Sun and the Earth" series, produced by the National Academy of Science; distributed by McGraw-Hill Films.)

Radio Astronomy in Australia (1968, color, 18 min). A discussion of radio astronomy techniques, with emphasis on charting the Sun and our Galaxy. (Produced by the Commonwealth Scientific and Industrial Research Organisation Film Unit; available from Indiana University.)

Solar Prominences (1948, black and white, silent, 11 min). Time-lapse studies of solar activity. A number of similar films are available from various sources. (Available from the University of Michigan.)

Film Loops

Solar Flares (1964, color, 4 min).

Solar Prominences (1964, color, 3 min). Both loops are time-lapse studies of solar activity. (Part of the "Space Science" series, produced and distributed by Bailey Film Associates.)

Slides

Slides of solar phenomena are available from Yerkes, Hale, and (to a lesser extent) Lick observatories, as well as from commercial suppliers.

SAMPLE EXAM QUESTIONS (MULTIPLE-CHOICE)

1. The region in the Sun from which the visible radiation comes is called the—
   a. limb
   b. photosphere
   c. corona
   d. chromosphere
   e. hydrogen convection zone
2. The granulation effect that can be observed on sharp pictures of the solar disk is the result of—
   a. limb-darkening
   b. the solar wind churning the surface
   c. rising and falling currents in the solar atmosphere
   d. differential rotation
   e. turbulence in the Earth's atmosphere
3. Astronomers know that the Sun is rotating from observations of—
   a. sunspots
   b. granulation
   c. limb-darkening
   d. eclipses
   e. the solar wind
4. Which layer of the Sun extends to the Earth?
   a. photosphere
   b. chromosphere
   c. hydrosphere
   d. solar wind
   e. more than one of the above
5. Which layer of the Sun is visible during a total eclipse of the

Sun?

a. photosphere
b. chromosphere
c. corona
d. more than one of the above
e. none of the above

6. The element helium was first discovered—
 a. by the Greeks
 b. in the absorption lines from the solar photosphere
 c. in the solar chromosphere during an eclipse
 d. in the laboratory after World War II
 e. by its radio emissions from the Sun
7. Great arcs of material extending along magnetic field lines from the chromosphere to the corona are called—
 a. prominences
 b. spicules
 c. sunspots
 d. helioarcs
 e. flares
8. The radio Sun is bigger than the optical Sun because—
 a. radio waves are bigger than light waves
 b. radio waves carry less energy than light waves
 c. radio waves interact strongly with electrons in the solar corona, which is much more transparent to light waves
 d. all radio waves in the Sun are generated in prominences that extend quite far above the photosphere
 e. disturbances in the Earth's ionosphere distort our "picture" of the radio Sun

Answers to Multiple-choice Questions

1. b 2. c 3. a 4. d 5. d 6. c 7. a 8. c

SAMPLE EXAM QUESTIONS (ESSAY)

1. Briefly list and describe the various layers near and above the Sun's surface. What sort of phenomena occur in each layer?
2. How do we know that the Sun is rotating?
3. How do the spectra of the chromosphere and the corona differ, and why?
4. Where would be the best place to search for emission lines from the solar corona? What sort of equipment would you need? Explain.
5. If prominences stick out into the hot corona, why are they visible?
6. An ecologically minded friend, who has not had the benefit of a good astronomy course, comes to you worried that the Sun is losing its atmosphere via the solar wind. What would you say to reassure him? (Be sure to explain how the solar wind is thought to be generated, as a part of your answer.)
7. Define and contrast (a) prominences, (b) flares, (c) sunspots, and (d) granulation.
8. An army general is worried about the Sun interfering with vital radio communications during a war. How valid is his worry? Explain.
9. An astronomer who studies the Sun has to write an application for a research grant to continue his work. The gossip around the

astronomy world has it that the astronomers on the government advisory committee that must judge his request are all interested in the study of other stars. If he is smart, what will our solar astronomer emphasize in the justification for his research? Explain briefly, with some examples.

CHAPTER 18
THE STRUCTURE OF A STAR

NOTE BY AUTHOR OF TEXT

This is another key chapter. Like Chapter 16, it not only develops a large part of how stars work and how we know it; it also brings in a lot of fundamental physics. Whereas Chapter 16 introduced the student to atoms, the present chapter acquaints him with atomic nuclei. Stellar structure and nuclear physics are closely interwoven in the flow of the chapter, which ends with a final emphasis on how the nature of atoms determines so much of the nature of the Universe. The material of the chapter also offers rich opportunities for discussions that should catch the interest of the student, from nuclear energy and the future of mankind to neutron stars and black holes.

—I.R.K.

SUMMARY OF KEY POINTS

1. Basic principles of stellar structure:
    a. a stable star is in balance—the compression pressure of gravity must equal the gas pressure that resists it;
    b. a star radiates energy from its surface—there must be a temperature difference to keep energy flowing outward;
    c. thus, if it is not to collapse and if it is to continue shining, a star must have an internal source of energy—this comes from nuclear fusion.
2. Fusion:
    a. because nuclei are positively charged, fusion can take place only when collision occurs with sufficient energy to overcome the electrical repulsion (i.e., fusion requires high temperatures);
    b. for most of its life (main sequence), a star will fuse hydrogen into helium. This can be done in two ways: the proton-proton chain, which dominates for lower main sequence stars; and the CNO cycle, which dominates for the higher ones.
3. Stages in the evolution of a star:
    a. a cloud of interstellar material begins to collapse, forming a protostar;
    b. (pre-main-sequence) under gravitational contraction, the star heats up until the onset of fusion at the center;
    c. (main sequence; the major part of the star's life) the star

converts H to He;

d. (post-main-sequence) H is exhausted in the core; fusion takes place in a surrounding shell; the core contracts, and the rest of the star expands; the star moves to the right in the HR diagram toward the red giant region; eventually, He fusion begins in the core;

e. (final states; after some mass loss in most cases)—

(1) white dwarf (for stars with final mass less than 1 to 1.5 $M_\odot$);

(2) neutron star (final mass from around 1 to a little more than 2 $M_\odot$);

(3) black hole (final mass significantly more than 2 $M_\odot$).

4. All the elements heavier than helium are believed to be built up only in stars. (Thus most of *our* atoms are products of stellar evolution.) The main processes of element building include—
   a. hydrogen fusion: 4 H → He [p-p chain or CNO cycle];
   b. triple-alpha process: 3 He → C;
   c. alpha process: add He to C, then to O, Ne, Mg, etc.;
   d. equilibrium (*e*) process: builds up all elements up to iron;
   e. slow (*s*) and rapid (*r*) neutron enrichment: builds up heavier elements—may require a supernova explosion.

Basic Terms Introduced in This Chapter

$E = mc^2$
Isotope
Deuteron
Fusion
Fission
CNO cycle
Proton-proton cycle
Protostar
Triple-alpha process
Degeneracy
Helium flash
Planetary nebula
Neutron star
Black hole
Nucleosynthesis
e-process
s-process
r-process

TOPIC FOR DISCUSSION AND CLASS ACTIVITIES

1. Students rarely think about stars as *evolving* objects. To make this point even more clearly, it can be instructive to draw a loose parallel between the stages in the life of a star and the stages in the life of a human being. For example:
   a. prenatal embryo = protostar;
   b. trauma of birth = onset of fusion;
   c. prolonged adolescence = main sequence;
   d. the vagaries of middle age = red giants, variability, mass loss;
   e. old age = white dwarfs;

f. death = supernova;
g. graveyard = black dwarfs, neutron stars, black holes.

2. Have students look up (or distribute or display) a list of the most abundant elements in the Universe. Compare this list with a list of the most abundant elements in the human body. Then discuss the statement by the late Harvard astronomer Harlow Shapley that we are all "children of the stars." (For one list of abundances, see "The Chemical Elements of Life," by E. Frieden, *Scientific American*, July 1972).

3. Have the students verify that at the Sun's prodigal rate of energy production, a Sun's worth of coal would last only a few thousand years. (Relevant data: Sun's mass = 2×10^{33} gm; energy from the burning of 1 gm of coal = 4×10^{11} ergs. Ignore the need for oxygen in the burning.)

4. Discuss the periodic table of elements and how it enabled scientists to predict the properties of then-undiscovered elements. We now know 92 long-lived elements; heavier elements may be formed during supernova explosions. Why don't we see these heavier elements? (A good set of articles on the transuranic elements by Glenn Seaborg et al. appeared in *Scientific American*, April 1950, December 1956, April 1963, and November 1970.)

5. This chapter points to a few of the problems involved with achieving controlled fusion in the laboratory. Students may enjoy looking into the progress being made in this area. *Science News*, *Physics Today*, and a number of other magazines regularly feature news about fusion research. See, for example:

Gough, W., and B. Eastlund. "Prospects of Fusion Power," *Scientific American*, February 1971.
Lubin, M., and A. Fraas. "Fusion by Laser," *Scientific American*, June 1971.
Thomsen, D. "Shiva, Argus and Co.," *Science News*, July 31, 1976.
Thomsen, D. "Mirror, Mirror in the Lab," *Science News*, August 21, 1976.

6. On p. 334 of the text, brief reference is made to the problem of the missing solar neutrinos. Students who want to look into this topic more closely have a good array of references to choose from:

Bahcall, J. "Neutrinos from the Sun," *Scientific American*, July 1969. [See also the taped interview with Dr. Bahcall listed under "Audio-visual Aids" below.]
Fowler, W. "The Case of the Missing Neutrinos," *Science Year 1974* (World Book Encyclopedia).
Reines, F. "Neutrino Astronomy," in *The New Universe* (Rand McNally, 1967-68).
Thomsen, D. "The Mystery of the Missing Neutrinos," *Science News*, September 25, 1971.
Thomsen, D. "The Solar Neutrino Mystery Deepens," *Science News*, March 10, 1973.
Thomsen, D. "Odd Particles Out," *Science News*, August 23-30, 1975.

For instructors who want more technical information about the neutrino problem, here are a few recent references:

Bahcall, J., and R. Davis. "Solar Neutrinos: A Scientific Puzzle," *Science*, January 23, 1976.

Barkat, Z. "Neutrino Processes in Stellar Interiors," in *Annual Reviews of Astronomy and Astrophysics*, vol. 13 (1975).

Trimble, V., and F. Reines. "The Solar Neutrino Problem—A Progress(?) Report," in *Reviews of Modern Physics*, vol. 45 (January 1973), p. 1.

Ulrich, R. "Solar Neutrinos and Variations in the Solar Luminosity," *Science*, November 14, 1975.

7. The mention of the positron as a product of the nuclear reactions in stars may be a good occasion for a more general discussion of antimatter, which many students seem to find fascinating. An interesting film called *Anti-matter* (1973, color, 11 min) was produced at UCLA and is distributed by American Educational Films. See also:

Alfvén, H. *Worlds-Antiworlds* (W. H. Freeman and Co., 1966).

Alfvén, H. "Antimatter and Cosmology," *Scientific American*, April 1967.

Alfvén, H., and A. Elvins. "Anti-matter, Quasi-stellar Objects, and the Evolution of Galaxies," *Science*, May 23, 1969. [More advanced.]

Burbidge, G., and F. Hoyle. "Anti-matter," *Scientific American*, April 1958.

Segré, E., and C. Wiegard. "The Antiproton," *Scientific American*, June 1956.

Thomsen, D. "Does the Universe Need Anti-matter?," *Science News*, March 31, 1973.

8. A number of science-fiction stories make reasonably reliable use of some of the concepts discussed in this chapter. Such stories can be a most effective way of motivating students to examine some of the abstract and difficult topics in stellar evolution in more detail. For example, see the following short stories:

Clarke, A. "The Star," in *The Nine Billion Names of God* (Signet paperback). [Features a supernova explosion.]

Niven, L. "Neutron Star," in *Neutron Star* (Ballantine paperback) or in *Where Do We Go from Here?* ed. I. Asimov (Fawcett Crest paperback). [Students may want to consider the plausibility of Niven's exploration of conditions around a neutron star. Note that the story was written before the discovery of pulsars and the consequent upsurge of interest in collapsed objects.]

Niven, L. "The Hole Man," in *A Hole In Space* (Ballantine paperback).

Niven, L. "The Borderland of Sol," in *Tales of Known Space* (Ballantine paperback). [Both this story and the preceding one use the notion of a quantum (or mini-) black hole, advanced by S. Hawking and others.]

Niven, L. "Flatlander," in *Neutron Star* (Ballantine paperback). [Involves a piece of antimatter that enters our Galaxy.]

9. A tremendous amount of public interest has focused in recent years on the black hole phenomenon. Many students will have read about black holes in the media and may request more information than is provided in the text. An extensive bibliography is therefore provided in the "Reading Materials" section below.

10. Someone once suggested that modern astronomy can be divided into the following halves: the Crab Nebula, and everything else. Although this statement may be somewhat exaggerated, it is certainly true that a fascinating array of phenomena can be seen at work in this stellar remnant. Instructors may find it useful to show the film *The Crab Nebula,* which outlines the steps in the discovery of the true nature and energy source of the nebula, and then to discuss how such supernova remnants fit into the scheme of stellar evolution, as well as their importance for the rest of astronomy.

READING MATERIALS FOR THE STUDENT

General Books

Baker, A. *Modern Physics and Antiphysics* (Addison-Wesley paperback, 1970). [A nice introduction to Einstein's theory of relativity.]

Calder, N. *Violent Universe* (Viking, 1969). [A little dated, but a nice introduction to modern astronomy, with a long section on stellar evolution.]

Jastrow, R. *Red Giants and White Dwarfs* (Signet paperback, 1971). [Elementary introduction to stellar evolution, and evolution in the cosmos in general.]

Meadows, A. *Stellar Evolution* (Pergamon Press, 1967). [A good descriptive summary, written before the discovery of pulsars.]

Shipman, H. *Black Holes, Quasars, and the Universe* (Houghton-Mifflin, 1976). [Superb introduction to several modern topics in astronomy; the first section deals with the endpoints of stellar evolution.]

Nucleosynthesis

Hack, M. "The Abundance of Helium in the Cosmos," *Sky and Telescope,* Sept.-Oct. 1972.

Kraft, R. "FG Sagittae: Rosetta Stone for Nucleosynthesis," *Sky and Telescope,* July 1974.

Mestel, L. "Why Does the Sun Shine?" in *The Emerging Universe,* eds. W. Saslaw and K. Jacobs (Univ. Press of Virginia, 1972).

Selbin, J. "The Origin of the Chemical Elements," *Journal of Chemical Education,* May-June 1973.

Stellar Evolution: Early Stages

Bok, B. "The Birth of Stars," *Scientific American,* August 1972.

Herbig, G. "The Youngest Stars," *Scientific American,* August 1967.

Strom, S., and K. Strom. "The Early Evolution of Stars," *Sky and Telescope,* May-June 1973.

Thomsen, D. "A Funny Thing Happened on the Way to the Main Sequence," *Science News,* January 10, 1976.

Stellar Evolution: Later Stages

See also references cited for "Pulsars" in Chapter 15, and for "Black Holes" below.

Gorenstein, P., and W. Tucker. "Supernova Remnants," *Scientific American,* July 1971.
Liller, W., and M. Liller. "Planetary Nebulae," *Scientific American,* April 1963.
Maran, S. "The Gum Nebula," *Scientific American,* December 1971.
Ruderman, M. "Solid Stars," *Scientific American,* February 1971.
Schramm, D., and W. Arnett. "Supernovae: The Origin of the Chemical Elements, Cosmic Rays, Neutron Stars and Maybe Even Black Holes," *Mercury,* May/June 1975.
Thorne, K. "Gravitational Collapse," *Scientific American,* November 1967.
Thorne, K. "The Death of a Star," in *Science Year 1968* (World Book Encyclopedia).

Black Holes

Block, D. "Black Holes and Their Astrophysical Implications," *Sky and Telescope,* July-August 1975.
Goldsmith, D., and D. Levy. *From the Black Hole to the Infinite Universe* (Holden-Day paperback, 1974).
"Interview with Steven Hawking," *Mercury,* Nov./Dec. 1975.
Kaufmann, W. *Relativity and Cosmology* (Harper & Row paperback, 1976). [The chapters on black holes are reprinted in *Mercury,* March-June 1974.]
Kaufmann, W. *Cosmic Frontiers of General Relativity* (Little, Brown, 1977).
Penrose, R. "Black Holes," *Scientific American,* May 1972.
Sullivan, W. "A Hole in the Sky," *The New York Times Magazine,* July 14, 1974.
Taylor, J. *Black Holes—The End of the Universe* (Random House, 1975).
Thomsen, D. "Black Holes: No Longer Hypothetical," *Science News,* January 13, 1974.
Thorne, K. "The Search for Black Holes," *Scientific American,* December 1974. [An excellent article.]
Wheeler, J. "Beyond the Black Hole," in *Science Year 1973* (World Book Encyclopedia).

READING MATERIALS FOR THE INSTRUCTOR

General Books

Baade, W. *The Evolution of Stars and Galaxies* (Harvard Univ. Press, 1963). [Edited transcripts of lectures by an astronomer who contributed enormously to these fields.]
Clayton, D. *Principles of Stellar Evolution and Nucleosynthesis* (McGraw-Hill, 1968). [A very popular graduate text in the field.]
Cox, J., and R. Giuli. *Principles of Stellar Structure,* 2 vols. (Gordon and Breach, 1968).
Fowler, W. *Nuclear Astrophysics* (American Philosophical Society, 1967). [Transcripts of Fowler's 1965 public lectures.]
Motz, L. *Astrophysics and Stellar Structure* (Ginn, 1970).
Shklovskii, I. *Supernovae* (Wiley Interscience, 1968).
Swihart, T. *Astrophysics and Stellar Astronomy* (Wiley, 1969).
Tayler, R. *Stars: Their Structure and Evolution* (Springer-Verlag,

1970).

General Article

Iben, I. "Stellar Evolution: Comparison of Theory with Observations," *Science*, vol. 155, p. 785 (1967).

Stellar Evolution: Early Stages

Cohen, M. "Star Formation and Early Evolution," *Mercury*, Sept./Oct. 1975. [Notes from an A.S.P. conference.]

Strom, S. "Star Formation and the Early Phases of Stellar Evolution," in *Frontiers of Astrophysics*, ed. E. Avrett (Harvard Univ. Press, 1976).

Strom, S., et al. "Young Stellar Objects and Dark Interstellar Clouds," in *Annual Reviews of Astronomy and Astrophysics*, vol. 13 (1975).

Nucleosynthesis

Ames, O. "Stars and Nuclei," *Physics Teacher*, April-May 1972.

Arnett, W. "Explosive Nucleosynthesis in Stars," in *Annual Reviews of Astronomy and Astrophysics*, vol. 11 (1973).

Bethe, H. "Energy Production in Stars," *Physics Today*, September 1968.

Clayton, D. "The Origin of the Elements," *Physics Today*, May 1969.

Fowler, W., et al. "Thermonuclear Reaction Rates II," in *Annual Reviews of Astronomy and Astrophysics*, vol. 13 (1975).

Reeves, H. "On the Origin of the Light Elements," in *Annual Reviews of Astronomy and Astrophysics*, vol. 12 (1974).

Schramm, D. "Nucleocosmochronology," in *Annual Reviews of Astronomy and Astrophysics*, vol. 12 (1974).

Trimble, V. "The Origin and Abundances of the Chemical Elements," in *Reviews of Modern Physics*, vol. 47 (October 1975), p. 877.

Unsold, A. "Stellar Abundances and the Origin of the Elements," *Science*, March 7, 1969.

Stellar Evolution: Later Stages

See also references listed for "Pulsars" in Chapter 15, and for "Black Holes" below.

Arnett, W. "Explosive Nucleosynthesis in Stars," in *Annual Reviews of Astronomy and Astrophysics*, vol. 11 (1973).

Cameron, A. "Endpoints of Stellar Evolution," in *Frontiers in Astrophysics*, ed. E. Avrett (Harvard Univ. Press, 1976).

Greenstein, G. "Prospects for Detecting Black-body X-rays from Neutron Stars," *Science*, August 9, 1974.

Gursky, H. "Neutron Stars, Black Holes and Supernovae," in *Frontiers in Astrophysics*, ed. E. Avrett (Harvard Univ. Press, 1976).

Iben, I. "Post-main Sequence Evolution of Single Stars," in *Annual Reviews of Astronomy and Astrophysics*, vol. 12 (1974).

Miller, J. "Planetary Nebulae," in *Annual Reviews of Astronomy and Astrophysics*, vol. 12 (1974).

Ostriker, J. "Recent Developments in the Theory of Degenerate White Dwarfs," in *Annual Reviews of Astronomy and Astrophysics*, vol. 9 (1971).

Salpeter, E. "Central Stars of Planetary Nebulae," in *Annual Reviews of Astronomy and Astrophysics*, vol. 9 (1971).

Sweigert, A. "Evolution of Red Giant Stars," *Physics Today*, Jan. 1976.

Tinsley, B. "What Stars Become Supernovae?" *Publication of the Astronomical Society of the Pacific*, December 1975.

Wheeler, J. C. "After the Supernova, What?" *American Scientist*, Jan.-

Feb. 1973.

Black Holes

Block, D. "General Relativity and Its Applications to Selected Astrophysical and Cosmological Problems," *Quarterly Journal of the Royal Astronomical Society*, vol. 15, p. 264 (1974).

Chandrasekhar, S. "The Black Hole in Astrophysics: The Origin of the Concept and Its Role," *Contemporary Physics*, January 1974.

DeWitt, C., and B. DeWitt (eds.). *Black Holes* (Gordon and Breach, 1973).

Eardley, D., and W. Press. "Astrophysical Processes Near Black Holes," in *Annual Reviews of Astronomy and Astrophysics*, vol. 13 (1975).

Gursky, H., and R. Ruffini (eds.). *Neutron Stars, Black Holes and X-ray Sources* (D. Reidel, 1975). [Includes reprints of important historical papers.]

Peters, P. "Black Holes," *American Scientist*, Sept./Oct. 1974.

Rees, M. "Black Holes," *Observatory*, August 1974.

Rees, M., R. Ruffini, and J. Wheeler. *Black Holes, Gravitational Waves and Cosmology* (Gordon and Breach, 1973).

Sciama, D. "The Physics and Astronomy of Black Holes," *Quarterly Journal of the Royal Astronomical Society*, March 1975.

AUDIO-VISUAL AIDS

Films

Anti-matter (1973, color, 11 min). History of the notion, followed by laboratory experiments and astronomical applications. (Distributed by American Educational Films.)

Birth and Death of a Star (1971, color, 27 min). John Wheeler narrates this visually dramatic film about stellar evolution. A number of astronomers are interviewed about their specialties; Jesse Greenstein speaks particularly eloquently about white dwarfs. (Distributed by Time-Life Films.)

The Crab Nebula (1970s, color, 52 min). This superb BBC-TV film treats the history of our understanding of the nebula as a most exciting detective story. Fine photography, well-edited interviews, and effective models all work together to create one of the best astronomy films ever made. (Distributed by Time-Life Films.)

The Origin of the Elements (1973, color). A lecture on nucleosynthesis by William Kaufmann, with some nice visual effects. (Produced and distributed by CRM Films.)

Universe on a Scratchpad (1964, color, 28 min). A somewhat disjointed film about theoretical astronomy at Goddard Institute, featuring Robert Jastrow and Patrick Thaddeus. Includes extensive discussion of the use of computers in this work; ends with a concise review of stellar evolution. (N.A.S.A. film HQ-164.)

Violent Universe (1969, black and white, 148 min). An ambitious, lengthy (5 reels!) survey of modern astronomy, with particular emphasis on stellar evolution, quasars, and cosmology. Features interviews with many prominent astronomers, as well as tours of the major observatories. A lot to see in one sitting. (Produced jointly by NET and BBC; distributed by Indiana University.)

Slides

Color and black-and-white slides of many of the objects discussed in this chapter are available from the Hale or Lick observatories, and from many secondary sources. For example, the Astronomical Society of the Pacific makes available to its members a set of color slides of the Orion, Crab, and Veil nebulae and the Pleiades Cluster.

Audio Tapes

The Birth of a Star (4 min) and *The Generation of Energy in Red Giant Stars* (9 min). William Fowler discusses topics in stellar evolution. (Distributed by Spring Green Multimedia.)

Blue Giants, White Dwarfs and Black Holes (60 min). June Engel interviews J. Percy and T. Bolton about stellar evolution. (Distributed by the Center for Cassette Studies.)

The Evolution of Stars (30 min) and *Relativity and the Death of Stars* (30 min). William Kaufmann discusses stellar evolution; both parts are included on the single tape ER-139. (Available from the Astronomical Society of the Pacific.)

Invisible Astronomy (20 min). Bruce Margon discusses observations in non-optical regions of the spectrum, including the discovery of a very hot white dwarf. (Astronomical Society of the Pacific tape ER-137.)

Solar Neutrinos: A Scientific Puzzle (30 min). John Bahcall is interviewed concerning the lack of neutrinos from our Sun. (Astronomical Society of the Pacific tape ER-145.)

Solving the Neutrino Puzzle (18 min). Philip Morrison talks about the solar neutrino question. (Distributed by Spring Green Multimedia.)

SAMPLE EXAM QUESTIONS (MULTIPLE-CHOICE)

1. A friend who does not have the vast new awareness you have gained from this course suggests that the Sun must be burning coal to shine as brightly as it does. Which of the following statements would *not* be part of the brilliant argument you would present to show him the error of his ways?
   a. Coal is made mainly of carbon; spectroscopic analysis of the Sun's radiation indicates that it contains relatively little carbon.
   b. If the Sun were burning coal, it could last only a few thousand years at the present rate of energy generation.
   c. The process of fusion in stars can *never* make carbon; all the carbon in the Universe is made on planets.
   d. The burning of coal on Earth requires oxygen; we see far too little oxygen in the Sun with our spectroscopes.
2. That same ignorant friend, impressed by your wisdom, asks you to tell him about fusion. After listening for a while, he breaks in: "Hey, I read somewhere that astronomers have observed water molecules in space. I bet the water forms by fusion of oxygen and hydrogen in the center of stars like the Sun!" What is your response?
   a. That's absolutely right! You're catching on, friend.
   b. There is absolutely no oxygen in any stars like the Sun.
   c. The fusion of oxygen and hydrogen into water can only happen in stars much older than the Sun, stars that have become red giants.
   d. The combination of oxygen and hydrogen to form water is not a fusion process.

3. Which of the following statements about the energy-generating process in the Sun is *true*?
   a. It can be duplicated efficiently on Earth with our present technology.
   b. Some mass is converted to energy.
   c. Hydrogen becomes helium in a single step.
   d. The forces involved are all electromagnetic.
4. Astronomers currently believe that elements heavier than iron are created in—
   a. the initial explosion of the Universe
   b. proton-proton chains
   c. degenerate white dwarfs
   d. supernova explosions
   e. They cannot be created at all.
5. A white dwarf, compared to a main sequence star of the *same mass*, would be—
   a. larger in radius
   b. smaller in radius
   c. younger
   d. lower in surface temperature
   e. less massive
6. According to our present theories, when our Sun "dies," what will it become?
   a. a white dwarf
   b. a black hole
   c. a neutron star
   d. a supernova
   e. a red giant
7. About 7 billion years ago, the heavier atoms now found in your body most probably were—
   a. in the star we call the Sun, undergoing fusion
   b. part of the planet Earth
   c. part of a previous-generation star, which would lose at least some of its mass before dying
   d. part of the hot fireball that created the Universe
   e. hanging around campus, trying to learn some astronomy
8. During the main sequence stage, the fuel that keeps a star like the Sun shining is—
   a. helium being coverted to carbon
   b. an enriched form of coal
   c. hydrogen being converted to helium
   d. the triple-alpha process
   e. a large supply of free neutrons
9. Astronomers claim that our Sun is a third-generation star. If a budding young astronomer wanted to prove this for himself, what sorts of observations would it be most useful for him to make?
   a. Measure the temperature of the Sun by looking at the distribution of different wavelengths in sunlight.
   b. Measure the distance to the Sun by using parallax and Kepler's law.
   c. Go above the Earth's atmosphere to see how much energy the Sun puts out in those wavelengths that do not reach the surface.
   d. Discover the composition of the Sun by measuring the discrete lines in the spectrum of sunlight.
   e. Measure the mass of the Sun by seeing its effect on bodies orbiting around it.

f. Measure the inside of the keyhole to the library where such information is kept by his professors.

10. The quantity that measures the resistance of stellar material to the flow of radiation is called the—
    a. temperature
    b. pressure
    c. opacity
    d. gravitational pull
    e. Come on, you can't fool me! Stellar material does not present any resistance to the flow of radiation.
11. Atoms of the same chemical element that have different atomic masses are called—
    a. radioactive
    b. deuterium
    c. nuclei
    d. fissionable
    e. isotopes
12. When a new star like our Sun is forming, what is the event that halts the initial gravitational contraction?
    a. the onset of fusion
    b. the helium flash
    c. the core becoming degenerate
    d. a supernova explosion
    e. the exhaustion of fuel supplies

Answers to Multiple-choice Questions

1. c 2. d 3. b 4. d 5. b 6. a 7. c 8. c 9. d
10. c 11. e 12. a

SAMPLE EXAM QUESTIONS (ESSAY)

1. In a spherical object like a star, gravity would be expected to pull everything in toward the center. Briefly explain how stars remain stable against the force of gravity.
2. Fission and fusion are both nuclear processes. How are they similar, and how are they different? Which of them is more important for stellar evolution, and why?
3. Briefly outline the steps in the evolution of a star with a mass like that of our Sun. Which stage in the evolution is the longest lasting, and why?
4. Briefly outline the steps in the evolution of a star ten times as massive as our Sun. How will the time required for this star's full life cycle compare to the time required for a star like the Sun? What is the reason for this difference?
5. What are the major processes for making heavier elements in stars? During what stages of a star's life is each process important?
6. Compare and contrast white dwarfs, neutron stars, and black holes. What type of star will end up as each type of remnant?
7. Briefly discuss the probable past history of (a) a hydrogen atom in the water you drink; (b) a carbon atom in your little finger, and (c) a gold atom in a piece of jewelry. In each case, begin the history with the origin of our Galaxy.
8. An eccentric old millionaire wishes to leave a large amount of money to help discover whether black holes exist. How would you counsel him on the writing of specific terms for his will? (No fair talking him out of it!)

CHAPTER 19
INTERSTELLAR MATERIAL

NOTE BY AUTHOR OF TEXT

The study of interstellar material is a wide and difficult field that occupies a large part of the concern of today's astronomers. The opening two paragraphs of the chapter set its theme: the nature of the interstellar material and its role in the astronomical picture. Some new physics is introduced, but the content of the chapter is almost completely astronomical. As always, the picture is built up directly from the observations; and many of the discussions are as open-ended as science itself.

—*I.R.K.*

SUMMARY OF KEY POINTS

Interstellar Material in General

1. There is almost as much material in the gas and dust of the interstellar medium as there is in the stars.
2. The interstellar gas has a composition like that of Population I stars; its density is around 1 particle/cc; its average temperature around 50°K. The gas is found in neutral regions, in low-density ionized regions, and in dense dust clouds in molecular form.
3. The interstellar dust is composed of grains with sizes around 10^{-5} cm; its composition is not clear; the dust composes about 1% of the interstellar material.

Emission Nebulae

1. Emission nebulae are clouds of gas surrounding the hottest stars.
2. Here, ultraviolet photons from the star ionize hydrogen in a hot spherical volume, called an HII ("H-two") region.
3. When electrons and ions recombine, a whole spectrum of lines is emitted, and the nebula is seen to glow. (Many of the transitions involved occur only under low-density conditions, and were thus termed "forbidden lines" in the lab.)

Interstellar Absorption Lines

1. The interstellar gas produces absorption lines that can be observed in the spectra of brighter, hot stars.
2. These lines will occur at different Doppler shifts from the lines of the star, indicating the motion of the gas clouds.

The Radio Lines of Hydrogen

1. A transition between the parallel and antiparallel spin states for electrons in the ground level of hydrogen results in the emission of a radio photon with wavelength of 21 cm.
2. Although such a transition has a low probability, there are so many hydrogen atoms in the interstellar medium that 21-cm radiation is relatively easy to observe.
3. Radio recombination lines are emitted as recombining electrons move between very high levels in the hydrogen atom.
4. Because most of the interstellar medium is hydrogen, these observations are crucial in understanding the material between the stars.

Interstellar Dust

1. The dust lies mainly along the galactic disk.
2. It dims and reddens the light of stars in the disk, an effect that must be corrected for before reliable distances can be measured.
3. The observed extinction curve implies that the dust must consist of particles only a few times smaller than the wavelength of visible light.
4. It produces a strong ultraviolet "bump" in the extinction curve, which indicates that some of the dust particles probably are graphite.
5. Two other types of grains may well also exist in the dust.
6. Grains are probably formed in the outer layers of cool stars. (There is strong evidence for shells of dust around some stars, glowing in the infrared.)
7. Spectral features in the infrared emission from circumstellar dust shells indicate the presence of silicates and silicon carbide.

The Interstellar Magnetic Field

1. The magnetic field aligns the interstellar dust grains perpendicular to the field. The elongated grains then polarize starlight parallel to the direction of the field.
2. The field seems to lie approximately along the spiral arms of the Galaxy.
3. Ionized gas in the presence of the field rotates the direction of polarization of radiation from extragalactic radio sources.
4. High-speed electrons spiraling about the field give off a broad spectrum of radio waves.

Cosmic Rays

1. Cosmic rays are high-energy electrons and ions moving through and around the Galaxy (energies range from 10^6 to 10^{20} eV).
2. The cosmic rays carry as much energy in themselves as that of all the rest of the interstellar medium.
3. The cosmic rays seem to be quite young compared to the Galaxy, but their origin is not yet understood.

Dust Clouds and Star Formation

1. In denser dust clouds, shielded from the ultraviolet radiation of stars, atoms join into molecules, many of whose radio lines have been observed.
2. A number of molecules that are the first building blocks of life have been found in these clouds, indicating that the chemical steps leading to life may not be as difficult as was once thought.

3. Large amounts of molecular hydrogen are found (by their ultraviolet emission) in the protected environment of dense clouds.
4. Dark globules within these clouds probably are the progenitors of protostars.

<u>Basic Terms Introduced in This Chapter</u>

Emission nebula
Lyman limit
Reflection nebula
HII region
Forbidden lines
Metastable energy levels
Hyperfine states
Radio recombination lines
Reddening
Extinction curve
Polarization
Electron volts
Cosmic rays

TOPICS FOR DISCUSSION AND CLASS ACTIVITIES

1. Nebulae (both emission and reflection types) are superb subjects for a project in astrophotography for students who own (or can be lent) the proper equipment. There are many good books on this subject, including *Outer Space Photography*, by H. Paul, which is available from Sky Publications. For the ultimate in professional astrophotography, students should look at the article and photos by J. Miller in *Scientific American*, October 1974.

2. The discovery of 21-cm radiation (and, in general, the development of radio astronomy) required the technical advances that were brought about by World War II. This is an interesting example of the interaction of science and technology and would make a nice paper topic for an historically minded student with a little technical background. The following sources provide a place to begin:

   Hey, J. *The Evolution of Radio Astronomy* (Neale Watson, 1973).
   Pfeiffer, J. *The Changing Universe* (Random House, 1956). [An engagingly written popular-level history of early radio astronomy.]
   Westerhout, G. "The Early History of Radio Astronomy," in "International Conference on Education in and History of Modern Astronomy," ed. R. Berendzen, *Annals of the New York Academy of Science*, August 25, 1972.

3. Several more complex molecules have been discovered in the interstellar medium since the text was written, and more are being found each year. *Science News* does a good job of reporting the new ones, and students can search through recent issues or the end-of-the-year index to compile an up-to-date list.

   The discussion of organic molecules in space always generates tremendous discussion in my classes. Accordingly, I have given a more extensive bibliography for this area in the "Reading

Materials," below.

4. In 1959, astrophysicist Fred Hoyle published a clever science-fiction novel, *The Black Cloud* (Signet paperback), which included the notion that intelligent life might evolve in the densest dust clouds. As our text discusses, recent discoveries about organic molecules make this proposal even more interesting. Students may want to look into possible connections between molecules in dust clouds and star formation in those regions. See, for example:

   Buhl, D., and C. Ponnamperuma. "Interstellar Molecules and the Origin of Life," *Space Life Sciences*, vol. 3, p. 157 (1971).
   Turner, B. "Interstellar Molecules," *Scientific American*, March 1973.

5. One interesting discovery that may bear on the relationship between interstellar molecules and the subsequent development of life on planets is the role of molecular *masers* in dispersing energy in protostar structures. Many of the references in "Reading Materials," below, address this point.

6. Although, as the text mentions, the 1420-MHz band is protected by international treaties, a number of other frequencies important for radio astronomy are not. An interesting but difficult research project might be to investigate how successful astronomers have been in protecting wavelengths in the past, and what they can do in the future. (This issue is very similar to the light-pollution problem mentioned in Topic #8 in Chapter 2.)

7. After reading this chapter, students should be encouraged to look back at Color Plate 8 with new appreciation. Note that half a million observations (plus incredible hours of computer and graduate-student time) went into this unique radio picture of our sky.

READING MATERIALS FOR THE STUDENT

See also the references cited under "Stellar Evolution: Early Stages" in Chapter 18.

General Books

Aller, L. *Atoms, Stars and Nebulae* (Harvard Univ. Press, 1971).

Page, T., and L. Page (eds.). *Stars and Clouds of the Milky Way* (Macmillan, 1969). [Reprints from *Sky and Telescope*.]

Verschuur, G. *The Invisible Universe* (Springer-Verlag, 1974). [A nice introduction to radio astronomy.]

General Articles

Barrett, A. "Radio Signals from Hydroxyl Radicals," *Scientific American*, December 1968.

Berge, B., and G. Seielstad. "The Magnetic Field of the Galaxy," *Scientific American*, June 1965.

Bok, B. "Birth of Stars," *Scientific American*, August 1972.

Gaustad, J. "The Composition of the Interstellar Dust," *Astronomical Society of the Pacific Leaflet* #483 (September 1969).

Goldberg, L. "Ultra-violet Astronomy," *Scientific American,* June 1969.
Heiles, C. "Cold Clouds and Hot Stars—An Evolutionary Sequence," *Astronomical Society of the Pacific Leaflet* #482 (August 1969).
Henry, R. "A Far Ultra-violet Look at Orion," *Sky and Telescope,* March 1972.
Herbig, G. "Interstellar Smog," *American Scientist,* March/April 1974.
Maran, S. "The Red Rectangle," *Natural History,* November 1975.
Miller, J. "The Structure of Emission Nebulae," *Scientific American,* October 1974.
Neugebauer, G., and E. Becklin. "The Brightest Infra-red Sources," *Scientific American,* April 1973.
Ney, E. "The Mysterious Egg Nebula," *Sky and Telescope,* January 1975.
Roberts, M. "Hydrogen in Galaxies," *Scientific American,* June 1963.
Snow, T. "A New Copernican Revolution in Astronomy: Ultra-violet Astronomy with the Satellite Copernicus," *Mercury,* March/April 1976.
Thomsen, D. "Where Stars Are Born," *Science News,* August 18-25, 1974.
Turner, B. "Interstellar Molecules," *Scientific American,* March 1973.
Vandervoort, P. "The Age of the Orion Nebula," *Scientific American,* February 1965.
Visvanathan, N. "Interstellar Polarization," *Astronomical Society of the Pacific Leaflet* #463 (January 1968).

Interstellar Molecules

Buhl, D. "Light Molecules and Dark Clouds," *Mercury,* Sept.-Dec. 1972.
Buhl, D. "Molecules and Evolution in the Galaxy," *Sky and Telescope,* March 1973.
Johnson, K., et al. "Microwave Celestial Water Vapor Sources," *Sky and Telescope,* August 1972.
McCray, R. "Molecules Between the Stars," *Natural History,* December 1974.
Milton, S. "Chemicals in the Sky," *Astronomical Society of the Pacific Leaflet* #507 (September 1971).
Snyder, L., and D. Buhl. "Molecules in the Interstellar Medium," *Sky and Telescope,* Nov.-Dec. 1970.
Thomsen, D. "A New Look at the Galaxy," *Science News,* December 4, 1971.
Turner, B. "Interstellar Molecules," *Scientific American,* March 1973.

Cosmic Rays

Aller, L. "The Chemical Composition of Cosmic Rays," *Sky and Telescope,* May-June 1972.
Burbidge, G. "The Origin of Cosmic Rays," *Scientific American,* August 1966.
Ginzburg, V. "The Astrophysics of Cosmic Rays," *Scientific American,* February 1969.

READING MATERIALS FOR THE INSTRUCTOR

See also references cited under "Stellar Evolution: Early Stages" in Chapter 18.

General

Aannestad, P., and E. Purcell. "Interstellar Grains," in *Annual Reviews of Astronomy and Astrophysics,* vol. 11 (1973).

Chaisson, E. "Gaseous Nebulae and Their Interstellar Environment," in *Frontiers of Astrophysics*, ed. E. Avrett (Harvard Univ. Press, 1976).

Dalgarno, A. "Chemistry of the Interstellar Medium," in *Frontiers of Astrophysics*, ed. E. Avrett (Harvard Univ. Press, 1976).

Dalgarno, A., and R. McCray. "Heating and Ionization of HI Regions," in *Annual Reviews of Astronomy and Astrophysics*, vol. 10 (1972).

Field, G., and A. Cameron (eds.). *The Dusty Universe* (Neale Watson, 1975). [Transcript of a 1973 conference on dust.]

Heiles, C. "Physical Conditions and Chemical Constitution of Dark Clouds," in *Annual Reviews of Astronomy and Astrophysics*, vol. 9 (1971).

Kaplan, S., and S. Pikel'ner. "Large-scale Dynamics of the Interstellar Medium," in *Annual Reviews of Astronomy and Astrophysics*, vol. 12 (1974).

Larson, R. "Processes in Collapsing Interstellar Clouds," in *Annual Reviews of Astronomy and Astrophysics*, vol. 11 (1973).

Middlehurst, B., and L. Aller (eds.). *Nebulae and Interstellar Matter* (Univ. of Chicago Press, 1968). [Vol. 7 of the "Stars and Stellar Systems" series.]

Neugebauer, G., et al. "Infrared Sources of Radiation," in *Annual Reviews of Astronomy and Astrophysics*, vol. 9 (1971).

Spitzer, L. *Diffuse Matter in Space* (Wiley, 1968).

Spitzer, L., and E. Jenkins. "Ultraviolet Studies of the Interstellar Gas", in *Annual Reviews of Astronomy and Astrophysics*, vol. 13 (1975).

Strom, S., et al. "Young Stellar Objects and Dark Interstellar Clouds," in *Annual Reviews of Astronomy and Astrophysics*, vol. 13 (1975).

Verschuur, G. "High Velocity Neutral Hydrogen," in *Annual Reviews of Astronomy and Astrophysics*, vol. 13 (1975)

Interstellar Molecules

Buhl, D., and C. Ponnamperuma. "Interstellar Molecules and the Origin of Life," *Space Life Sciences*, vol. 3 (1971), p. 157.

Cook, A. "Molecules in Space," *Quarterly Journal of the Royal Astronomical Society*, March 1975.

Dalgarno, A. "Chemistry of the Interstellar Medium," in *Frontiers of Astrophysics*, ed. E. Avrett (Harvard Univ. Press, 1976).

Gammon, R. "Chemistry Between the Stars." [N.A.S.A. pamphlet for teachers, planned for publication in late 1976.]

Litvak, M. "Coherent Molecular Radiation," in *Annual Reviews of Astronomy and Astrophysics*, vol. 12 (1974).

Moran, J. "Radio Observations of Interstellar Masers," in *Frontiers of Astrophysics*, ed. E. Avrett (Harvard Univ. Press, 1976).

Rank, M., et al. "Interstellar Molecules and Dense Clouds," *Science*, December 10, 1971.

Snyder, L. "Chemistry Between the Stars," in *The Emerging Universe*, ed. W. Saslaw and K. Jacobs (Univ. Press of Virginia, 1972).

Solomon, P. "Interstellar Molecules," *Physics Today*, March 1973.

Spitzer, L. "Hydrogen Molecules in Interstellar Space," *Quarterly Journal of the Royal Astronomical Society*, June 1976.

Zuckerman, B., and P. Palmer. "Radio Radiation from Interstellar Molecules," in *Annual Reviews of Astronomy and Astrophysics*, vol. 12 (1974).

Cosmic Rays

Cowsik, R., and P. Price. "Origin of Cosmic Rays," *Physics Today*, October 1971.
Israel, M., et al. "Ultraheavy Cosmic Rays," *Physics Today*, May 1975.
Meyer, P., et al. "Cosmic Rays: Astronomy with Energetic Particles," *Physics Today*, October 1971.
Wentzel, D. "Cosmic Ray Propagation in the Galaxy: Collective Effects," in *Annual Reviews of Astronomy and Astrophysics*, vol. 12 (1974).

AUDIO-VISUAL AIDS

Films

Cosmic Rays (1960, color, 27 min). Review of the history and current status of the study of cosmic rays. Features a brief discussion of atomic, nuclear, and particle physics and of theories about the origin of these particles. (Part of the "Planet Earth" series, produced by the National Academy of Science; distributed by McGraw-Hill Films.)

Fields of Space (1970, color, 19 min). Discussion of the interplanetary and interstellar media and their interaction with magnetic fields. (Produced by the National Film Board of Canada; distributed by McGraw-Hill Films.)

A Radio View of the Universe (1967, color, 28 min). See description in Chapter 20.

Slides

Emission nebulae are among the most beautiful of astronomical objects. Hale, Lick, and Yerkes observatories, as well as the commercial suppliers, all make slides of the objects discussed in the text (and others) available, many in color.

Audio Tapes

The Development of 21-cm Radio Astronomy (10 min). A brief talk by H. van de Hulst. (Available from Spring Green Multimedia.)

Invisible Astronomy (20 min). An interview with ultraviolet and X-ray astronomer Bruce Margon. (Astronomical Society of the Pacific tape ER-137.)

SAMPLE EXAM QUESTIONS (MULTIPLE-CHOICE)

1. Full and empty are very relative terms. In a room we term empty, there are still about 10^{19} particles per cubic centimeter. In the interstellar medium, the average particle density is close to—
   a. 0 particle/cc
   b. 1 particle/cc
   c. 100 particles/cc
   d. 10^4 particles/cc
   e. 10^{15} particles/cc
2. An HII ("H-two") region is—
   a. a cloud of hydrogen molecules
   b. an atmosphere of hydrogen around stars like our Sun
   c. a source of 21-cm radiation

d. a sphere of ionized gas
e. a cloud in which water molecules have been found

3. When astronomers observe "forbidden lines," they know the radiation is coming from—
   a. gas of very low density
   b. very cold hydrogen gas
   c. dust clouds
   d. the central regions of hot stars
   e. regions of strong magnetic fields
   f. a hidden tape recorder which was monitoring conversations in the White House
4. The 21-cm radiation is the result of transitions between—
   a. metastable states of oxygen
   b. molecular bands of graphite
   c. the two spin states of the hydrogen ground level
   d. free and bound states for electrons in hydrogen
   e. none of the above
5. Most of the interstellar medium consists of—
   a. dust particles
   b. molecules containing hydrogen
   c. hydrogen atoms
   d. cosmic rays
   e. free electrons
6. The amount of reddening of starlight by dust can be measured because dust does not affect the star's—
   a. apparent magnitude
   b. color
   c. apparent brightness
   d. spectral type
   e. blue wavelengths
7. For which of the following substances is there *not* astronomical evidence in the interstellar dust?
   a. graphite
   b. fluorocarbons
   c. silicates
   d. silicon carbide
   e. All of the above are observed in dust "out there."
8. Which of the following is a way to detect the presence of the interstellar magnetic field?
   a. 21-cm radiation
   b. forbidden lines
   c. polarization of starlight
   d. recombination radiation
   e. the twinkling of pulsars
9. In denser clouds, most of the hydrogen is believed to be in the form of—
   a. separate atoms
   b. hydrogen molecules
   c. dust grains
   d. water (ice)
   e. none of the above

<u>Answers to Multiple-choice Questions</u>

1. b   2. d   3. a   4. c   5. c   6. d   7. b   8. c   9. b

SAMPLE EXAM QUESTIONS (ESSAY)

1. Most of the time, interstellar material is invisible. Describe two different conditions under which this material becomes visible, and discuss how it is then detected.
2. Why would you not expect to detect HII regions and emission nebulae around G-type stars like our Sun?
3. Suppose an astronomer observes interstellar absorption lines of calcium originating in several clouds between us and a star that also contains calcium. Why would he *not* worry about confusing effects of calcium in the star's atmosphere with effects of calcium in the interstellar clouds?
4. Contrast the means by which radio astronomers detect cold and hot hydrogen gas.
5. What's forbidden about "forbidden lines"?
6. A certain Midwestern senator charges that, of the money being allocated for astronomy, too much is being spent on radio telescopes with equipment to observe hydrogen. How would you defend the astronomical community against these charges of "hydrogen chauvinism"?
7. Why is interstellar dust the astronomer's curse, and how do we get around the problem it causes?
8. Why is the sky blue? Why is the sunset red?
9. The space between the stars is very sparsely populated, with relatively few atomic collisions. How then do astronomers believe that entire dust grains form?
10. What are the three forms in which interstellar hydrogen exists? Under what conditions is each form found?
11. Discuss three ways to convince a skeptic that our galaxy has a magnetic field.
12. In 1959, an imaginative astrophysicist named Fred Hoyle wrote *The Black Cloud*, a science-fiction novel in which he proposed that some form of life may have started in dense interstellar dust clouds. Discuss recent evidence that makes Hoyle's proposal a little more plausible than it seemed when he wrote the book.

CHAPTER 20
THE MILKY WAY, OUR HOME GALAXY

NOTE BY AUTHOR OF TEXT

This is a long and densely packed chapter. In some sense, the study of the Milky Way, and also of other stellar systems, is the synthesis toward which the previous chapters have been building. Yet much of the chapter is concerned with the properties of the Milky Way itself, and in this sense it is a bridge to the extragalactic chapters that follow.

—I.R.K.

TEACHING NOTES

Many students will find some of the concepts in this chapter a little confusing. In particular, the details of obtaining radial velocities from 21-cm profiles seems to give nonscience students a great deal of trouble. It helps to emphasize at each point why such measurements are important (perhaps through elucidation of the "general on a fog-bound battlefield" analogy on p. 385).

The paragraph on the local density (p. 397) may be, if you'll pardon the expression, a bit dense for many students. A review of the luminosity function (p. 251) and the mass-luminosity relation (pp. 260 and 328) as well as a clear class discussion may be needed here.

Students who read the text and look at the diagrams carefully may be a little confused by one point concerning differential rotation. The discussions on pp. 392 and 404, as well as Figure 20.7, emphasize that the objects toward the Galactic center rotates faster than we do, and that objects farther out rotate more slowly. These statements *seem* to apply to the whole Galaxy. However, according to the diagram on p. 394, the inner parts of the Galaxy are rotating more slowly.

The reason for this discrepancy is that the inner part of the Galaxy seems to be in solid-body-type rotation, whereas the outer parts are moving in Keplerian-type orbits. It should be brought out in class discussion that the statements on pp. 392, 393, and 404 apply only to "neighborhoods" near the Sun.

SUMMARY OF KEY POINTS

General

1. Our Galaxy consists of 200 billion stars in a disk 30,000 pc in diameter, rotating around the center with periods of hundreds of millions of years.
2. Population I stars make up most of the disk; Population II stars are distributed in a more extensive halo.

Discovering the Structure of the Galaxy

1. The band of the Milky Way indicates that the stars form a disk and are more concentrated toward the center; these observations are even better in the infrared, where dust obscuration is reduced.
2. "Windows" in the dust help us to see further inward, using RR Lyrae stars as distance indicators.
3. The center of the system of globular clusters, which outline the Galactic halo, should be the center of the Galaxy. Its distance is about 10 kiloparsecs.
4. By observing other galaxies similar to ours, we surmise that the Milky Way must have spiral arms too.
5. The hydrogen distribution can be mapped from Doppler shifts in the 21-cm observations, once the Galactic rotation curve is found.
6. The arms are dotted by HII regions around bright O and B stars, which can be discerned for some distance away.

Rotation of the Galaxy

1. The Sun moves around the Galactic center at a speed of 250 km/sec and with a period of 250 million years.
2. The Galaxy is in differential rotation; the rotation curve can be determined from the velocity profiles of the 21-cm radiation.

Galactic Dynamics

1. The rotation curve enables us to calculate the total mass of the Galaxy (about 200 billion times the mass of the Sun).
2. The amount of 21-cm radiation tells us the amount of interstellar material, which is a few percent of the mass in stars.
3. We can determine the density of our local region from the oscillations of stars perpendicular to the Galactic disk.
4. Visible stars account for 50% of the local density; 10 to 20% is interstellar atomic hydrogen; the rest may be molecules, dwarf stars, or may represent uncertainties in our figures.
5. As a result of differential rotation, the spiral arms should wind up. The fact that the spiral structure maintains itself indicates that the spiral arms must be produced by a wave of compression maintained in the gas, perhaps by gravitational instability.
6. As the wave of compression moves through, it may trigger formation of stars.

The Galactic Halo

1. The halo is outlined by the globular clusters (and other tracers).
2. Its stars can be distinguished in the local neighborhood by their large proper motions. (About 1% of the nearby stars belong to the halo population.)
3. Halo stars are Population II, old and metal-deficient.
4. Objects with greater metal abundance show a flatter distribution with respect to the Galactic disk.

Stellar Groups

1. *Globular clusters:* rich, spherical groups in the halo; about 125 are known.
2. *Open (galactic) clusters:* smaller groups, less densely packed, located in the disk. Over 700 are known; there must be very many more blocked by dust clouds.
3. *Stellar associations:* groups of stars that, though born together, separate in a few million years. It may be that all stars are born in such associations.
4. Each of these groups consists of stars born at the same time out of the same material; they remain together by gravitational attraction, but dwindle as stars "evaporate" from the group. The time scale for disruption is proportional to the cluster size.

Basic Terms Introduced in This Chapter

Differential rotation
Rotation curve
Gravitational instability
Open clusters
Stellar associations

TOPICS FOR DISCUSSION AND CLASS ACTIVITIES

1. Although I am in principle very much in favor of teaching students to reason in the metric system, some numbers simply cry out for translation into units they are more familiar with. For example, students will be most impressed to learn that the speed of our rotation around the center of the Galaxy quoted in the text corresponds to about 550,000 mph. As an exercise, they may want to compare this to the orbital speed of the Earth around the Sun (more than 60,000 mph) or the speed of light (670,000,000 mph). For maximum inpact, let *them* do the calculations, however.

2. Although our text concentrates on our modern-day understanding of the Galaxy, there is an interesting history to the discovery that we live in one spiral-shaped grouping of stars among an immensity of others. The books cited in the first subsection of "Reading Materials for the Student," below, cover much of this history.

   In particular, students may want to look into the career and thought of William Herschel, who did some fascinating and pioneering work on the structure of the Milky Way. See, for example, *William Herschel and the Construction of the Universe,* by M. Hoskin (Neale Watson, 1963).

3. The late-nineteenth-century controversies about the size and uniqueness of the Milky Way culminated in 1920 in a public debate between Harlow Shapley and Heber Curtis. This debate (which also bears on the material in Chapter 21) would make an excellent topic for a student report, panel discussion, or even restaging. See, for example:

   Hetherington, N. "The Shapley-Curtis Debate," *Astronomical Society of the Pacific Leaflet* #490 (April 1970).
   Struve, O. "A Historic Debate About the Universe," *Sky and Telescope,* May 1969.

4. Although our text eschews mentioning modern astronomers by name, one whose career may make an interesting topic for a report is Harlow Shapley. See his autobiography, *Through Rugged Ways to the Stars* (Scribner's, 1969), and consult:

   Bok, B. "Harlow Shapley: Cosmographer and Humanitarian," *Sky and Telescope*, December 1972.
   Bok, B. "Harlow Shapley and the Discovery of the Center of Our Galaxy," in *The Heritage of Copernicus*, ed. J. Neyman (MIT Press, 1974).
   Focus on Harlow Shapley. A tape-recorded interview with Shapley; see "Audio-visual Aids," below.

5. The presence of a paper on Shapley in a volume commemorating Copernicus (Topic #4) serves to remind us how much the work of Shapley and others establishing our place in the Galaxy followed the Copernican spirit and left mankind even farther from any central place in the Universe. Students often enjoy a brief discussion on the possible philosophical implications of astronomical research, and this may be a good place to have one.

6. Many of the photos for this chapter are identified with M (Messier) numbers. Students are always amused to learn that Charles Messier (1730-1817) was a dedicated comet hunter, whose famous and enduring list of the most spectacular nonstellar objects was compiled so as to avoid confusing these annoying nebulae and other objects with the comets that were his true passion.

7. On November 16, 1974, at the occasion of the dedication of the new surface for the 1,000-foot radio telescope at Arecibo, a message was beamed in the direction of the globular cluster M13. In traveling the 24,000 light years (approximately 8,000 parsecs) to this cluster, the message will spread until it covers the width of the cluster. For more about the "text," see *Science News*, November 23, 1974.

READING MATERIALS FOR THE STUDENT

Historically Oriented Books

Berendzen, R., et al. *Man Discovers the Galaxies: Case Studies on the Development of Modern Astronomy* (Neale Watson, 1976).
Jaki, S. *The Milky Way: An Elusive Road for Science* (Neale Watson, 1972).
Kopal, Z. *Widening Horizons: Man's Quest to Understand the Structure of the Universe* (Taplinger, 1970).
Pfeiffer, J. *The Changing Universe* (Random House, 1956). [History of early radio astronomy.]
Struve, O., and V. Zebergs. *Astronomy in the 20th Century* (Macmillan, 1962). [See Chapter 19.]
Whitney, C. *The Discovery of Our Galaxy* (A. Knopf, 1971).

Historically Oriented Articles

Bok, B. "Harlow Shapley and the Discovery of the Center of Our Galaxy," in *The Heritage of Copernicus*, ed. J. Neyman (MIT Press,

1974).
Ewen, H. "Radio Waves from Interstellar Hydrogen," *Scientific American,* December 1953. [Ewen was the codiscoverer of 21-cm radiation from space.]
Oort, J. "The Development of Our Insight into the Structure of the Galaxy Between 1920 and 1940," in "International Conference on Education in and History of Modern Astronomy," ed. R. Berendzen, *Annals of the New York Academy of Science,* August 25, 1972.
Weaver, H. "Steps Toward Understanding the Large Scale Structure of the Milky Way," *Mercury,* Sept./Oct. 1975 and Jan./Feb. 1976.
Westerhout, G. "The Early History of Radio Astronomy," in "International Conference on Education in and History of Modern Astronomy," ed. R. Berendzen, *Annals of the New York Academy of Science,* August 25, 1972.

General Books

Bok, B., and P. Bok. *The Milky Way,* 4th ed. (Harvard Univ. Press, 1974).
Page, T., and L. Page. *Stars and Clouds of the Milky Way* (Macmillan, 1968).
Schatzman, E. *The Structure of the Universe* (McGraw-Hill paperback, 1968). [See Section 2.]
Verschuur, G. *The Invisible Universe* (Springer-Verlag paperback, 1974). [A nice introduction to radio astronomy.]

General Articles

Arp, H. "On the Origin of Spiral Arms in Galaxies," *Sky and Telescope,* December 1969.
Bok, B. "An International Discussion of Milky Way Research," *Sky and Telescope,* July 1963.
Bok, B. "The Spiral Structure of Our Galaxy," *Sky and Telescope,* December 1969 and January 1970.
Garrison, R. "Stellar Associations," *Astronomical Society of the Pacific Leaflet* #465 (March 1968).
"Gould's Belt: An Expanding Group of Stars," *Sky and Telescope,* February 1968.
Iben, I. "Globular Cluster Stars," *Scientific American,* July 1970.
King, I. "The Dynamics of Star Clusters," *Sky and Telescope,* March 1971.
"Our Galaxy As Mapped with RR Lyrae Stars," *Sky and Telescope,* February 1968.
Roberts, M. "Hydrogen in Galaxies," *Scientific American,* June 1963.
Sanders, R., and G. Wrixon. "The Center of the Galaxy," *Scientific American,* April 1974.
Woolley, R. "The Mechanism of Our Galaxy," in *The New Universe* (Rand McNally, 1967-68).

READING MATERIALS FOR THE INSTRUCTOR

Historically Oriented Materials

Fernie, J. "The Historical Quest for the Nature of the Spiral Nebulae," *Publications of the Astronomical Society of the Pacific,* December 1970.
Hey, J. *The Evolution of Radio Astronomy* (Neale Watson, 1973).
van de Kamp, P., et al. "Five Articles Honoring Harlow Shapley on His

Eightieth Birthday," *Publications of the Astronomical Society of the Pacific,* October and December 1965.

In addition, many of the books and articles recommended for students (preceding section) merit the instructor's attention.

General Books

Blaauw, A., and M. Schmidt (eds.). *Galactic Structure* (Univ. of Chicago Press, 1965). [Volume 5 of the "Stars and Stellar Systems" series.]

Mavridis, L., (ed.). *Structure and Evolution of the Galaxy* (Springer-Verlag, 1971).

Mihalas, D., and P. Routly. *Galactic Astronomy* (W. H. Freeman and Co., 1968).

General Articles

Aarseth, S., and M. Lecar. "Computer Simulations of Stellar Systems," in *Annual Reviews of Astronomy and Astrophysics,* vol. 13 (1975).

Bertojo, M. "Isotopic Abundances and Their Variations Within the Galaxy," *Science,* May 10, 1974.

Bok, B. "Updating Galactic Spiral Structure," *American Scientist,* Nov./Dec. 1972.

Cameron, A., and J. Truran. "The Chemical Evolution of the Galaxy," *Journal of the Royal Astronomical Society of Canada,* February 1971.

King, I. "Stellar Populations," *Publications of the Astronomical Society of the Pacific,* August 1971.

Roberts, M. "Interstellar Hydrogen in Galaxies," *Science,* February 1, 1974.

Schwarzschild, M. "Stellar Evolution in Globular Clusters," *Quarterly Journal of the Royal Astronomical Society,* March 1970.

Shu, F. "Spiral Structure, Dust Clouds, Star Formation," *American Scientist,* Sept./Oct. 1973.

van den Bergh, S. "Stellar Populations in Galaxies," in *Annual Reviews of Astronomy and Astrophysics,* vol. 13 (1975).

Woolley, R. "The Stars and the Structure of the Galaxy," *Quarterly Journal of the Royal Astronomical Society,* December 1970.

AUDIO-VISUAL AIDS

Films

Exploring the Milky Way (1967, color, 29 min). See description in Chapter 15.

Galaxies (1971, color, 12 min). Discussion of our Galaxy and others, as well as how they are studied. (Produced by Haberstroh Studios; distributed by McGraw-Hill Films.)

Radio Astronomy in Australia (1968, color, 18 min). See description in Chapter 17.

Radio View of the Universe (1967, color, 28 min). Morton Roberts narrates this A.A.S. film about the use of radio telescopes in studying our Galaxy and others. (Distributed by Modern Learning Aids.)

What Are Stars Made Of? (1960, color, 16 min). See description in Chapter 16. Includes discussion of abundance differences between globular-cluster and solar-type stars.

Slides

Slides of galaxies like the Milky Way, and of star clusters and associations, are available from all three of the major observatory collections, as well as from commercial supply houses.

Audio Tapes

Focus on Harlow Shapley (28 min). An interview with the late Harvard astronomer. (Available from the Center for Cassette Studies.)

SAMPLE EXAM QUESTIONS (MULTIPLE-CHOICE)

1. According to modern observations, what is the approximate distance to the center of the Galaxy?
   a. 93 million miles
   b. 10 parsecs
   c. 10 kiloparsecs
   d. 30 kiloparsecs
   e. We still have no idea of that distance.
2. Our knowledge of the total mass of the Galaxy comes from—
   a. RR Lyrae stars
   b. globular clusters
   c. the Galactic magnetic field
   d. the Galactic rotation wave
   e. "windows" between large dust clouds
3. Our knowledge of the local density of material in this part of the Galaxy comes from—
   a. RR Lyrae stars
   b. the motions of stars perpendicular to the Galactic plane
   c. radio radiation from molecules in local dust clouds
   d. interstellar reddening
   e. clustering of stars
4. Which objects are particularly useful for finding the outline of the spiral arms of the Galaxy?
   a. HII regions
   b. dust clouds with molecules in them
   c. globular clusters
   d. the Galactic halo
   e. bright population II stars
5. The Galactic halo is—
   a. the strong glow of the central region of our Galaxy
   b. the strong radio emission we observe from the Orion arm in which we find ourselves
   c. a thinly spread distribution of stars all around and beyond the Galactic disk
   d. a chain of open clusters near the center of the Galaxy
   e. an important religious symbol
6. Why are globular clusters easier to find than galactic clusters?
   a. Globular clusters are all very close to our local neighborhood.
   b. There are many more globular clusters in the Galaxy than galactic clusters.
   c. Globular clusters are extremely strong sources of radio radiation.
   d. Globular clusters are large and lie outside the plane of the Galaxy.

e. Globular clusters formed relatively recently and therefore contain many extremely bright stars.

7. Globular clusters can survive longer than the average galactic cluster because—
   a. globular clusters generally contain more stars
   b. globular clusters generally contain fewer stars
   c. globular clusters were born at the beginning of the Galaxy
   d. globular clusters replace lost stars with new ones
   e. globular clusters are metal deficient

Answers to Multiple-choice Questions

1. c 2. d 3. b 4. a 5. c 6. d 7. a

SAMPLE EXAM QUESTIONS (ESSAY)

1. In the sixteenth century, Copernicus created a revolution in mankind's perception of its place in the cosmos by displacing the Earth from a central place in the Solar System. How have our studies of the Milky Way Galaxy continued this Copernican trend?
2. After a long voyage in suspended animation, an astronaut of the far future wakes up in a (to him) unknown part of the Galaxy. Assuming he has all sorts of astronomical instruments at his disposal, how would he determine the direction of and distance to the Galactic center?
3. How is it that we can determine the mass of the Galaxy from the inside? How can we determine what fraction of that mass is in interstellar material?
4. How do we know that we live in a spiral galaxy? How can we determine where the main spiral arms are?
5. How does the Galaxy maintain its spiral structure? Why do the arms not wind up or unravel?
6. Give a brief but informative guided tour of the Galactic halo.
7. Why would you not expect to find a small open cluster as old as the Galaxy?

CHAPTER 21
GALAXIES: THE STELLAR CONTINENTS

NOTE BY AUTHOR OF TEXT

Extragalactic astronomy is one of the busiest fields of modern research; almost half of the total concern of astronomers goes into the subjects covered in Chapters 21 through 23. The study of galaxies is closely related to that of the Milky Way and stellar populations. Our knowledge of galaxies is still young and incomplete, and I have had to make a special effort to give a rational picture of their nature and workings.

--I.R.K.

TEACHING NOTES

Many of the ideas behind the distance scale can be quite abstract and can be confusing for some students. It may help to talk first about the general concept of "standard candles," a set of light sources whose intrinsic brightness does not vary. Many of the extragalactic distance methods are just searches for good standard candles.

Some instructors may prefer to spend a little more time on the velocity-distance relation (Hubble's law), and to discuss its history and development. (See Topic #8 in "Topics for Discussion," below.)

For those courses whose time is running short (and this late in the term, whose isn't?), the sections on populations in galaxies could be skipped without impairing the flow of the chapter.

The section on dynamics also includes some complex arguments and should be discussed with care, accompanied perhaps by a demonstration or film loop on angular-momentum conservation.

SUMMARY OF KEY POINTS

Galaxy Types

1. Ellipticals have a smooth distribution of light, contain only old stars, and have very little interstellar material. They range from E0s, which are round and show very little rotation, to E7s, which rotate faster and are therefore somewhat flattened.
2. Spirals show great variations in properties. S0 is a transition class between ellipticals and spirals. Moving from Sa's to Sd's, spirals show less and less of a nucleus and more spiral arms, as well as more interstellar material and young stars. Some have a

bar across the central region.

3. Irregulars have 10% or more of their material in interstellar gas and show many younger stars. But they contain older Type II stars as well.

The Distance Scale

1. The period-luminosity relation for cepheid variables lets us find distances out to 4 or 5 megaparsecs.
2. Comparing the absolute magnitudes of the brightest stars, the diameters of the largest HII regions, or the brightnesses of globular clusters gets us out to 10 megaparsecs.
3. Assuming that the brightest galaxies in any cluster have the same absolute magnitude can allow us to measure distances out to a billion parsecs.
4. For all distant galaxies, we can use the redshift and the velocity-distance relation to get distances.

Populations in Galaxies

1. Galaxies vary quite a bit in luminosity and size.
2. Dwarf galaxies, which are hard to detect, may be more numerous than large ones.
3. The dominant stellar population of the Universe is old Population I.
4. Metal abundances seem to decrease with distance from the center in many galaxies.

Dynamics of Galaxies

1. Masses of galaxies can be estimated from their internal motions; they show a large range.
2. M/L (the mass-luminosity ratio) is generally the same for galaxies with similar populations, but it varies as the mixture of populations changes.
3. In a protogalaxy, internal motions can support a gaseous system against gravity for only a very short time because of excitation and radiation as gas atoms or clouds collide. For a system that has already condensed into stars, there are no collisions and essentially no dissipation of energy.
4. Therefore, elliptical or halo systems must have formed stars early, and thus maintained the rounded shape of the protogalaxy cloud. Disk systems must have remained gaseous for a longer time, collapsed as energy was dissipated, and turned into a spinning flattened disk.
5. Because stars mix but gas clumps, galaxies with larger proportions of interstellar material may find it more difficult to form smooth structures. This could explain the difference between spirals and irregulars.

Groups and Clusters of Galaxies

1. Most galaxies are members of groups; some are grouped in rich clusters.
2. In our Local Group, the Milky Way and M31 dominate, but there are a number of smaller spirals, irregulars, and ellipticals. There may also be many dim dwarf ellipticals, but each is so small that their contribution to the total mass is not very significant.
3. In rich clusters (with diameters on the order of megaparsecs), a galaxy can take a billion years or more to cross the cluster. Galaxies in dense clusters have very little interstellar material

and are thus mainly ellipticals or S0s.

4. In those clusters that have achieved an equilibrium between individual motions and gravitational attraction, we can estimate the mass of the cluster. In rich clusters, this estimate is about ten times more than the observed mass of the galaxies. This "missing mass" could be in dwarf intergalactic stars, or rocks, or black holes, or extended halos of nonluminous objects around the galaxies in the cluster.

Basic Terms Introduced in This Chapter

Elliptical, spiral, and irregular galaxies
Barred spirals
Velocity-distance relation for galaxies
Mass-luminosity ratio for galaxies
Protogalaxy
The Local Group
The missing mass

TOPICS FOR DISCUSSION AND CLASS ACTIVITIES

1. The whole topic of other galaxies can sometimes be a very difficult one for students to come to terms with. (One instructor tells the story that a young student dropped the course when she learned that there are other galaxies. "The scale of the Milky Way is bad enough to grasp," she said, "but that there should be others . . .") On the other end of the spectrum, many students have trouble getting beyond thinking of galaxies as pretty, if fuzzy, pictures in the text. During initial discussions of galaxies, I like to introduce the notion that, in some of those galaxies, there may be astronomy classes who are looking at slides or textbook pictures of the Milky Way.

2. Every college library should purchase a copy of the beautiful *Hubble Atlas of Galaxies*, ed. A. Sandage (available from California Institute of Technology Bookstore). Students should be encouraged to browse at their leisure.

3. To illustrate the rather complex dynamical arguments in this chapter, every effort should be made to use diagrams and demonstrations. For example, to demonstrate the conservation of angular momentum, have a volunteer stand on a rotation table with a copy of the text in each extended hand. Then have him bring his arms in.

4. Occasionally, new members of the Local Group are discovered, sometimes with some "hoopla" in the general media. One example is the Maffei galaxies (see *Sky and Telescope*, March 1971). Another is the recent discovery of a galaxy (nicknamed "Snickers") by its radio emission (see "Noted in the Current Journals," *Mercury*, March/April 1976). Students might try to look up and evaluate newspaper and magazine coverage of such discoveries.

5. An interesting topic for class discussion might be to evaluate the effects of varying galactic types on the prospects for life in a galaxy. It should be stressed that such judgments (tentative

as they are) could only be made on the basis of "life as we know it."

6. Edwin Hubble's observations of a cepheid in M31 finally settled the controversy (which also formed in important part of the Shapley-Curtis debate; see Topic #3 in Chapter 20) concerning the existence of separate "island universes" like the Milky Way. The history of this problem makes fascinating reading. See, for example:

   Berendzen, R., et al. *Man Discovers the Galaxies* (Neale Watson, 1976).
   Glyn-Jones, K. *The Search for the Nebulae* (Science History Publications, 1975).
   Goldsmith, D. "Edwin Hubble and the Universe Outside Our Galaxy," in *The Heritage of Copernicus*, ed. J. Neyman (MIT Press, 1974).
   Jaki, S. *The Milky Way: An Elusive Road for Science* (Neale Watson, 1972).
   Struve, O., and V. Zebergs. *Astronomy in the 20th Century* (Macmillan, 1962).

7. Edwin Hubble is one of the few astronomers to appear on the cover of *Time* magazine. Students might enjoy finding that issue (February 9, 1948) and looking into the life of one of this century's most able observers. A series of lectures he gave about his own work has been collected in a volume called *The Realm of the Nebulae* and republished by Dover Books.

8. Hubble was also responsible for the velocity-distance relation that is discussed in this chapter (and again in Chapter 23). See Hubble's book (Topic #7) and the historical references in Topic #6 for the origin and early history of the diagram. A recent article to bring the student up to date is "A New Look at the Hubble Diagram," by K. Lang and G. Mumford, *Sky and Telescope*, February 1976.

9. It is worth reminding students, as they read about extragalactic distance measurements in this chapter, of the comments about the importance of distance measurements made at the beginning of Chapter 12. In addition, students can be reminded that each further step in measuring cosmic distances depends very much on all the previous steps. This is often termed the "distance pyramid," because the stability of each upper slab depends on the correct positioning of the slabs below. A good concise summary of all the relevant distance methods is given by H. Shipman in his new book, *Black Holes, Quasars and the Universe* (Houghton-Mifflin paperback, 1976).

READING MATERIALS FOR THE STUDENT

Books

Berendzen, R., et al. *Man Discovers the Galaxies* (Neale Watson, 1976).

Hodge, P. *Galaxies and Cosmology* (McGraw-Hill paperback, 1966).

Hodge, P. *Concepts of the Universe* (McGraw-Hill paperback, 1969).
Hubble, E. *The Realm of the Nebulae* (1936; reprinted as Dover paperback). [Classic essays by the "father of extragalactic astronomy."]
Page, T., and L. Page *Beyond the Milky Way* (Macmillan, 1969). [Reprints from *Sky and Telescope*.]
Schatzman, E. *The Structure of the Universe* (McGraw-Hill paperback, 1968). [See Section 4.]
Shapley, H., and P. Hodge. *Galaxies*, 3rd ed. (Harvard Univ. Press, 1972).

Articles

Arp, H. "The Evolution of Galaxies," *Scientific American*, January 1963.
Arp, H. "On the Origin of Arms in Spiral Galaxies," *Sky and Telescope*, December 1969.
Bok, B. "The Large Clouds of Magellan," *Scientific American*, January 1964.
Hodge, P. "Dwarf Galaxies," *Scientific American*, May 1964.
Hodge, P. "Some Current Studies of Galaxies," *Sky and Telescope*, July 1972.
Irwin, J. "Some Current Problems Concerning Galaxies," *Sky and Telescope*, November 1973.
"Maffei 1: A New Nearby Galaxy," *Sky and Telescope*, March 1971.
Margon, B. "The Missing Mass," *Mercury*, Jan./Feb. 1975.
Page, T. "The Evolution of Galaxies," *Sky and Telescope*, Jan.-Feb. 1965.
Rees, M., and J. Silk. "The Origin of Galaxies," *Scientific American*, June 1970.
Roberts, M. "Hydrogen in Galaxies," *Scientific American*, June 1963.
Roberts, M. "Galaxies: Landmarks of the Universe," in *The Emerging Universe*, ed. W. Saslaw and K. Jacobs (Univ. Press of Virginia, 1972).
Rubin, V. "The Dynamics of the Andromeda Nebula," *Scientific American*, June 1973.
Thomsen, D. "How the Universe Got Its Lumps," *Science News*, February 26, 1972.
Toomre, A., and J. Toomre. "Violent Tides Between Galaxies," *Scientific American*, December 1973.

READING MATERIALS FOR THE INSTRUCTOR

Books

Gribbin, J. *Galaxy Formation: A Personal View* (Wiley, 1976).
Jacobs, K. *Extragalactic Astronomy*. [N.A.S.A. pamphlet for high-school science teachers, to be published in late 1976.]
McVittie, G. *Problems of Extragalactic Research* (Macmillan, 1962).
Sandage, A., et al. *Galaxies and the Universe* (Univ. of Chicago Press, 1976). [Vol. 9 of the "Stars and Stellar Systems" series.]

Articles

Davis, M. "Galaxies and Cosmology," in *Frontiers of Astrophysics*, ed. E. Avrett (Harvard Univ. Press, 1976).
Hodge, P. "Dwarf Galaxies," in *Annual Reviews of Astronomy and Astrophysics*, vol. 9 (1971).

Oort, J. "Galaxies and the Universe," *Science*, December 25, 1970.
Peimbert, M. "Chemical Composition of Extra-galactic Nebulae," in *Annual Reviews of Astronomy and Astrophysics*, vol. 13 (1975).
Roberts, M. "Interstellar Hydrogen in Galaxies," *Science*, February 1, 1974.
van den Bergh, S. "Stellar Populations in Galaxies," in *Annual Reviews of Astronomy and Astrophysics*, vol. 13 (1975).

AUDIO-VISUAL AIDS

Films

Galaxies (1971, color, 12 min). See description in Chapter 20.
Radio View of the Universe (1967, color, 28 min). See description in Chapter 20.
The Realm of the Galaxies (1959, color, 18 min). Features Allan Sandage in a night at Palomar observatory, with emphasis on measuring distances to galaxies. (Part of the "Horizons of Science" series; distributed by Educational Testing Service.)

Slides

Nice slides and prints of galaxies, both in color and black and white, can be purchased from the Hale, Lick, and Yerkes observatories, or from many commerical suppliers.

SAMPLE EXAM QUESTIONS (MULTIPLE-CHOICE)

1. The type of spiral galaxy that is most similar to an elliptical is--
   a. SBa
   b. Sb
   c. SO
   d. Sa
   e. Sc
2. Which type of galaxy contains the greatest amount of interstellar material?
   a. elliptical
   b. spiral
   c. barred spiral
   d. irregular
   e. All types contain about the same amount of interstellar material.
3. Dr. Hyron Spinrad, a California astronomer, recently found a galaxy that is about 2.5 billion parsecs away. To measure this distance, Dr. Spinrad must have used--
   a. cepheids
   b. the velocity-distance relation
   c. the absolute magnitude of the brightest star in the galaxy
   d. the diameter of the largest HII region in the galaxy
   e. a lucky guess
4. To measure the distance to the Andromeda galaxy, our nearest independent neighbor galaxy, astronomers would be best off using--
   a. cepheids
   b. the velocity-distance relation

c. the total absolute magnitude of the galaxy
d. the magnitude of the brightest star in the galaxy
e. a radar beam

5. When an astronomer uses the velocity-distance relation to find the distance of a new galaxy, he must first measure that galaxy's--
   a. total absolute magnitude
   b. Doppler shift
   c. spectral type
   d. hydrogen content
   e. M/L
6. The most common types of stars in the Universe are--
   a. old Population I
   b. young Population I
   c. old Population II
   d. young Population II
   e. low-metal-abundance stars
7. Which region would have the lowest M/L ratio?
   a. a giant elliptical galaxy
   b. a globular cluster
   c. a spiral arm dominated by young Population I stars
   d. a region with a good mixture of Population I and Population II stars
   e. All of the above have about the same M/L.
8. According to current thinking, why didn't the material in an elliptical galaxy collapse into a disk like the plane of the Milky Way?
   a. Ellipticals have strong magnetic fields that prevent collapse.
   b. Ellipticals still have large amounts of interstellar matter that interferes with collapse.
   c. All ellipticals start out as very small clouds, so that the conservation of angular momentum can't flatten them much.
   d. A huge number of supernovae and HII regions keep an elliptical galaxy extended.
   e. In an elliptical, the material probably formed stars early, before a great deal of energy was dissipated.
9. There are two subclusters in our Local Group, one centered on the Milky Way, the other on--
   a. the Magellanic Clouds
   b. M31 (Andromeda)
   c. the Sombrero galaxy
   d. a tight group of dwarf ellipticals
   e. You can't fool me! The entire Local Group is centered on the Milky Way.
10. The "missing mass" problem refers to--
   a. the absence of interstellar material from elliptical galaxies
   b. the absence of neutrinos from the centers of stars like our Sun
   c. the scarcity of dwarf ellipticals in the Local Group
   d. the discrepancy between the dynamic and luminous mass of rich clusters
   e. the absence of spiral arms from irregular galaxies

<u>Answers to Multiple-choice Questions</u>

1. c 2. e 3. b 4. a 5. b 6. a 7. c 8. e 9. b 10. d

SAMPLE EXAM QUESTIONS (ESSAY)

1. Someone suggests that, because ellipticals contain few young stars, spirals more, and irregulars even more, there must be an evolutionary sequence in which galaxies are first irregular, then spiral, then elliptical. What evidence could you cite against this proposal?
2. I have a large radio telescope, tuned to a range of wavelengths around 21 cm. I point the telescope at three equidistant galaxies (an elliptical, a spiral, and an irregular). From which galaxy am I likely to get the strongest signal, and why?
3. What would be the most reliable way to measure the distance to each of the following? (a) a nearby galaxy within 3 megaparsecs; (b) a galaxy about 8 megaparsecs away; (c) a cluster of galaxies over 100 megaparsecs away; (d) an extremely distant galaxy not part of a cluster.
4. List three characteristics that differ among galaxies. Discuss briefly what sort of range of these characteristics each type of galaxy can display.
5. Why do the mass/luminosity ratios of various galaxies differ?
6. Contrast the dynamic history of a spiral galaxy with a very flat disk to that of an E0-type elliptical.
7. Why do galaxies in dense clusters show so little interstellar gas and dust? (Who's cleaning up?)
8. Where's the "missing mass" (in rich clusters of galaxies)?

CHAPTER 22
COSMIC VIOLENCE: FROM RADIO GALAXIES TO QUASARS

NOTE BY AUTHOR OF TEXT

This chapter contains a lot of exciting material, but I have dealt with it rather briefly. The reason is that only phenomenological descriptions are available; the theory and the understanding are not. I have therefore tried to convey the mystery of radio galaxies and quasars without devoting excessive attention to theoretical studies that are conflicting and probably irrelevant.

--I.R.K.

TEACHING NOTES

For those instructors who want to have a much longer discussion of these topics, extensive references are included in "Reading Materials," below.

SUMMARY OF KEY POINTS

General

1. The development of radio telescopes and interferometers made possible an accurate determination of radio-source positions.
2. The optical counterparts of many strong sources turned out to be galaxies or quasars.

Radio Galaxies

1. Radio galaxies give off synchrotron radiation in huge quantities, indicating explosions releasing energies equivalent to the lifetime energy production of many millions of stars in some cases.
2. They often have symmetric double sources, as much as 100 kiloparsecs apart.
3. Estimates for their lifetimes are a few million years.
4. When significant amounts of gas are present, this gas is often grossly disrupted by the explosion.
5. The center of our Galaxy also shows some radio emission, but much weaker than the strong sources.

Galaxies with Bright Nuclei

1. Galaxies with bright nuclei are generally called N galaxies.
2. They have nuclei whose brightness dominates the image of the

galaxy.
3. One subclass, the Seyfert galaxies, show strong wide emission lines from a small region at the center.
4. Such galaxies seem to be intermediate between normal galaxies and quasars.

Quasars

1. Quasars show very large redshifts. If these are interpreted as caused by velocities of recession, the quasars must be extremely distant and luminous.
2. Although their light and radio output is tremendous, fluctuation time-scales indicate very small sizes (fractions of a parsec).
3. In addition to continuum radiation, they give off bright emission lines typical of excited interstellar nebulae.
4. There are other schools of thought about quasars, which propose that
   a. quasars are local objects shot out at high velocity from our Galaxy, or
   b. quasars are local objects whose redshift is caused by an unknown mechanism.
5. Quasars are difficult to identify, and we have found only a small fraction of the estimated 100,000 or more that may exist.
6. Quasars appear to be more common at larger distances than they are close to us, indicating that they were more common in the past.

Basic Terms Introduced in This Chapter

Interferometer
Synchrotron radiation
Seyfert galaxy
N galaxy
Quasar

TOPICS FOR DISCUSSION AND CLASS ACTIVITIES

1. The controversy over the interpretation of quasar redshifts recently generated a public debate similar to the Shapley-Curtis debate mentioned in the "Topics" for the preceding two chapters. Transcripts of this debate, together with papers that each side feels support its arguments, can be found in a marvelous paper-bound volume, *The Redshift Controversy*, by G. Field, H. Arp, and J. Bahcall (Benjamin, 1973).

2. Halton Arp, the Hale observatory astronomer who has been a central figure on the "local" side of this controversy, is interviewed at some length in *Mercury*, Nov./Dec. 1974.

3. Students might enjoy studying a number of current references on quasars and then staging a debate of their own.

4. Quasars are an excellent topic for pointing out that many problems in science remain controversial and maddeningly unsolved for years and years. Students may want to do a report comparing the debate over the need for new physics (or at least new insights) for quasars to other such episodes in the history of

science, such as the photoelectric experiment, the problem with black-body curves, the debate between Kelvin and Darwin on the time-scale available for evolution on Earth, etc.

5. Many unusual (even bizarre) theories have been proposed for the energy source of quasars. The references in the next section can fill in interested students on the details. They should keep large grains of interstellar salt with them at all times while doing this reading.

6. H. Alfvén and others who have been looking into the question of antimatter in the Universe look upon quasars with special interest. See, for example, "Anti-matter, Quasi-stellar Objects and the Evolution of Galaxies," by H. Alfvén and A. Elvins, *Science*, May 23, 1969.

7. An interesting recent science-fiction novel by astronomer Fred Hoyle, *The Inferno* (Harper & Row, 1973), uses a local quasar/active-galaxy phenomenon as its central focus of action. (For more on quasars and cosmology according to Hoyle, see his book, *From Stonehenge to Modern Cosmology*, W. H. Freeman and Co., 1972.) The entire problem of quasars (together with other trivial questions like the origin of the "big bang") is neatly solved in a very clever and prize-winning science-fiction novel by biochemist Isaac Asimov, *The Gods Themselves* (Fawcett Crest paperback, 1972).

8. Two pieces of music have been based on quasars. Clermont Pepin's "Symphony #3" is subtitled "Quasars" and features a movement called "CTA 102." It is available on a Canadian record. Ernst Widmer's "Quasars" was premiered in 1970 and was recorded in Brazil. (The rock group, The Byrds, also recorded a song called "CTA 102.")

READING MATERIALS FOR THE STUDENT

Books

Calder, N. *Violent Universe* (Viking paperback, 1969). [Section II is devoted to active galaxies and quasars.]

Hodge, P. *The Revolution in Astronomy* (Holiday House paperback, 1970). [See Chapters 3 and 4.]

Kaufmann, W. *Relativity and Cosmology* (Harper & Row paperback, 1976). [See Chapter 9.]

Shipman, H. *Black Holes, Quasars and the Universe* (Houghton-Mifflin paperback, 1976). [A superb, up-to-date review.]

Verschuur, G. *The Invisible Universe* (Springer-Verlag paperback, 1974). [See Chapters 10 through 13.]

Articles

Arp, H. "The Quasar Controversy" [an interview], *Mercury*, Nov./Dec. 1974.

Burbidge, G., and F. Hoyle. "The Problems of Quasi-stellar Objects," *Scientific American*, December 1966.

Burbidge, M., and R. Lynds. "Absorption Lines in Quasi-stellar Objects," *Scientific American*, December 1970.

"Discoveries in NGC 5128," *Sky and Telescope*, August 1975.
Green, L. "Quasars Six Years Later," *Sky and Telescope*, May 1969.
Heeschen, D. "Radio Galaxies," *Scientific American*, March 1962.
Hodge, P. "Some Current Studies of Galaxies," *Sky and Telescope*, July 1972.
Irwin, J. "Some Current Problems Concerning Galaxies," *Sky and Telescope*, November 1973.
Jastrow, R. "The Quasar Controversy," *Natural History*, May 1974.
"Light Variations in Quasars," *Sky and Telescope*, July 1969.
McCrea, W. "Energy Sources in Galaxies and Quasars," in *McGraw-Hill Yearbook of Science and Technology* (McGraw-Hill, 1972).
Pacini, F., and M. Rees. "Rotation in High-energy Astrophysics," *Scientific American*, February 1973.
Sandage, A. "Exploding Galaxies," *Scientific American*, November 1964.
Sanders, R., and G. Wrixon. "The Center of the Galaxy," *Scientific American*, April 1974.
Schmidt, M., and F. Bello. "The Evolution of Quasars," *Scientific American*, May 1971.
Strom, R., et al. "Giant Radio Galaxies," *Scientific American*, August 1975.
Stull, M. "Two Puzzling Objects," *Sky and Telescope*, April 1973.
Toomre, A., and J. Toomre. "Violent Tides Between Galaxies," *Scientific American*, December 1973.
Vorontsov-Velyaminov, B. "Quasars as the Pre-clusters of Galaxies," *Astronomical Society of the Pacific Leaflet* #485 (November 1969).
Weymann, R. "Seyfert Galaxies," *Scientific American*, January 1969.

READING MATERIALS FOR THE INSTRUCTOR

Books

Burbidge, G., and E. Burbidge. *Quasi-stellar Objects* (W. H. Freeman and Co., 1967). [Slightly dated.]
Evans, D. *External Galaxies and Quasi-stellar Objects* [I.A.U. Symposium #44] (Springer Verlag, 1972).
Field, G., H. Arp, and J. Bahcall. *The Redshift Controversy* (Benjamin, 1973).
Sandage, A., et al. (eds.). *Galaxies and the Universe* (Univ. of Chicago Press, 1976). [Vol. 9 of the "Stars and Stellar Systems" series.]
Sciama, D. *Modern Cosmology* (Cambridge Univ. Press, 1971). [See Chapters 4 through 7.]

Articles

Arp, H. "Observational Paradoxes in Extragalactic Astronomy," *Science*, December 17, 1971.
Brecher, K. "Active Galaxies," in *Frontiers in Astrophysics*, ed. E. Avrett (Harvard Univ. Press, 1976).
Colgate, S. "Quasistellar Objects and Seyfert Galaxies," *Physics Today*, January 1969.
Kellermann, K. "Extragalactic Radio Sources," *Physics Today*, October 1973.
Metz, W. "Double Radio Sources: Energetic Evidence That Galaxies Remember," *Science*, June 27, 1975.
Morrison, P. "Resolving the Mystery of the Quasars," *Physics Today*, March 1973.

Schmidt, M. "Quasars and Cosmology," *Observatory*, December 1971.
van den Bergh, S. "The Classification of Active Galaxies," *Journal of the Royal Astronomical Society of Canada*, June 1975.

AUDIO-VISUAL AIDS

Films

To the Edge of the Universe (1969, color, 23 min). Documentary about the construction of a Canadian radio telescope and how it is used to study quasars. (Produced by the National Film Board of Canada; distributed by McGraw-Hill Films.)

Violent Universe (1969, black and white, 148 min). See description in Chapter 18.

Slides

Slides of a few of the best-known active galaxies can be obtained from the Hale observatories.

Audio Tapes

Comets, Quasars and Black Holes (24 min). A discussion by Lloyd Motz. (Available from the Center for Cassette Studies.)

The Discovery and Controversy of Quasi-stellar Objects (18 min). A discussion by Maarten Schmidt. (Available from Spring Green Multimedia.)

Quasars (20 min). An interview with Dr. Larry Caroff. (Available from the Astronomical Society of the Pacific.)

Space, Void and Cosmos: Thomas Gold (30 min). The noted astrophysicist discusses quasars. (Available from the Center for Cassette Studies.)

SAMPLE EXAM QUESTIONS (MULTIPLE-CHOICE)

1. To determine the position of a radio source accurately, astronomers use--
   a. a large radio antenna
   b. a radio spectrometer
   c. an interferometer
   d. a satellite that can go above the Earth's atmosphere
   e. the FBI
2. The radiation emitted by radio galaxies results from the process of--
   a. recombination after ionization
   b. synchrotron radiation
   c. hydrogen-atom electrons in the ground state changing the direction of their spin
   d. molecules in dark clouds changing their states of rotation and vibration
   e. nuclear fusion
3. Quasars appear to be starlike points in the sky. The real size of a quasar can be estimated from--
   a. its emission lines
   b. its redshift
   c. its absolute luminosity

    d. the period of its fluctuations
    e. 21-cm radio emission

4. Two astronomers, both rather drunk, meet at a party and immediately begin arguing about quasars. One claims that there is good evidence that quasars are as far away as their redshifts indicate; the other claims that there is evidence that quasars are relatively close by. How would you resolve this debate?
    a. One of them is more drunk than the other; the evidence supports only *one* side in this debate.
    b. Quasars have only been predicted by theory; it is not at all clear that they have ever been observed.
    c. The problem of quasar redshifts and distances is one of the unsolved mysteries of modern science, so everyone might as well have another drink.
    d. Quasar distances are not a problem because the energy-emitting regions of quasars seem to be huge.
    e. A *sober* look at the facts indicates that quasars can be found everywhere; in our local spiral arm, in other galaxies, *and* very far away.
5. The fact that we see more quasars at larger distances than close to us implies that--
    a. quasars are thrown out by the center of the Milky Way
    b. quasars must be a fraction of a parsec or less in diameter
    c. quasars were more common in the past
    d. quasars are moving away from us with tremendous redshifts
    e. quasars show a large ultraviolet excess

Answers to Multiple-choice Questions

1. c   2. b   3. d   4. c   5. c

SAMPLE EXAM QUESTIONS (ESSAY)

1. How do astronomers estimate the age of strong radio galaxies?
2. What makes Seyfert galaxies peculiar?
3. Briefly discuss a pair of observations about quasars that appear to be contradictory.
4. How do the non-Establishment schools of thought about quasars differ from the Establishment school? Summarize arguments on both sides.
5. You're walking along on campus when a shadowy figure lurking in a doorway offers to sell you a hot new list of 10,000 quasars. Why would you be smart to be doubtful about such an offer?
6. How do quasars show that we live in an evolving Universe?

CHAPTER 23
THE UNIVERSE

NOTE BY AUTHOR OF TEXT

Cosmology catches the interest of nearly every student. It is full of strange and exciting ideas, but it is also full of paradoxes and pitfalls. I have tried to face these directly and to treat each of them as honestly as possible. Cosmology is also a vehicle for introducing some important ideas about the basic nature of mathematics and physics, and I have tried particularly to develop this kind of insight in the treatments of curved space and of general relativity.

--I.R.K.

TEACHING NOTES

In this chapter, more than any other, the ideas tend to "boggle" students' minds. Care must be taken in guiding the student gradually into these elusive concepts and in emphasizing the difficulty of the observations.

I tend to stress throughout the discussion of cosmology the fundamental difference between the subject of this chapter and all the others: while there are other planets, other stars, and other galaxies we can use for comparison, there is only *one* Universe. In cosmology, there can be no experiments with controls.

Students will have enormous difficulty with the notion of curved space. Positive curvature can be made much more clear by the use of a balloon painted with specks. (I use a durable weather balloon, available from Edmund Scientific Company, blown up using the "exhaust" pipe of the vacuum cleaner used with our physics-lab air track.)

Finally, let me strongly recommend planning the course in such a way that, if time does run short, it isn't the fascinating and mind-expanding notions of this chapter that are given short shrift. But unless you are one of those rare individuals who will be reading this entire manual before they begin the course, this advice is probably too late to do any good. Ah, well, *next time* you teach the course . . .

SUMMARY OF KEY POINTS

1. The Cosmological Principle is the assumption that the Universe would look the same from any observing point.
2. The redshift of a galaxy (and therefore its recessional velocity)

is proportional to its distance, for any observer.

3. The separation of the galaxies implies a time, something like 20 billion years ago, when they were all together. This time-scale is on the order of the ages of the oldest stars.
4. The beginning of the expansion, from a singular state of infinite density, is termed the "big bang" and marks the beginning of time.
5. The existence of the "big bang" event was recently confirmed by the discovery of a 3° black-body radiation background left from the time the radiation and matter in the Universe decoupled, now redshifted tremendously.
6. During the first few minutes after the big bang, when temperatures were in the billions of degrees, neutrons and protons fused into hydrogen, deuterium, and helium in a proportion determined by the density and temperature. All heavier atoms were made later by stars.
7. The Universe may obey various geometrical rules. It could have--
   a. zero curvature, which means it would be flat and would follow Euclidean geometry, where (1) the sum of the angles of a triangle equals 180°; (2) through a point not on a given line, there is only one line that never meets the given line; and (3) the volume of a sphere increases as the radius cubed.
   b. positive curvature, where (1) the sum of the angles of a triangle is more than 180°; (2) all lines intersect; and (3) the volume of a sphere increases less rapidly than the radius cubed.
   c. negative curvature, where (1) the sum of the angles of a triangle is less than 180°; (2) through a point not on a given line, there are many lines that never intersect the given line; and (3) the volume of a sphere increases more rapidly than the radius cubed.

   If galaxies are distributed uniformly, they should in principle enable us to measure how volume increases with radius. In practice, however, such observations are at present too difficult and uncertain.
8. There are three scenarios about the future dynamics of the Universe corresponding to these three geometries:
   a. the density of the Universe is "critical," just enough to halt the expansion at time infinity;
   b. the density of the Universe is larger than the critical value, so the Universe is closed and will fall back together again;
   c. the density of the Universe is smaller than the critical value, so the Universe is open and will continue expanding forever.
9. To decide among these models, we can use the following techniques.
   a. Summing the density of all known material in the Universe. This gives only a few percent of the critical value, but our survey may miss any undetected intergalactic matter.
   b. Examining redshifts of distant galaxies to compare present-day and early expansion. However, if galaxies evolve they cannot be used as "standard candles" for this comparison.
10. General relativity connects curvature and dynamics by showing that gravitation, in a sense, manifests itself in the curvature of space-time.

Basic Terms Introduced in This Chapter

Cosmology
The Cosmological Principle
Big bang
Singularity
Steady-state theory
Non-Euclidean geometry
Critical density

TOPICS FOR DISCUSSION AND CLASS ACTIVITIES

1. What good is cosmology? Why are we interested in the questions raised in this chapter? (Be prepared for spirited discussion.)

2. In the 1600s, Archbishop Ussher (of Ireland) apparently counted the "begats" in the book of Genesis and the Gospel of Matthew, added up the lifetimes, and was able to conclude that the Universe was created on a Sunday in October, 4004 B.C. This estimate remained as a popular figure for the age of the Universe for some time. It might be a nice exercise at the end of an astronomy course to bring together all the arguments the class can muster to show that the Universe is older than that.

3. Students who have trouble with some of the non-Euclidean geometry in this chapter (and even those who don't) might enjoy reading the classic little geometric allegory *Flatland* by Edwin Abbott (Barnes and Noble paperback, 1963). An intriguing little film based on the book is listed in "Audio-visual Aids," below. A similar book extending the story by one dimension is *Sphereland* by D. Burger (Crowell, 1965).

4. As the text emphasizes, the observations at present are too ambiguous and difficult to permit any final decision among the cosmological models. And the interpretation of the observations can also change as we gain a better understanding of the complex phenomena involved. Nevertheless, for a well-written progress report on the current status of cosmological observations, students should read "Will the Universe Expand Forever?" by J. Gott et al., *Scientific American*, March 1976.

5. In the article cited in Topic #4, the authors point out that the saddle shape is not the best analogy for negative curvature in space. Rather, they point to a geometric object called the *pseudosphere*, which they feel is a better representation. An intriguing print by M. C. Escher (called "Circle Limit IV"), which is published with the article, shows some of the properties of a pseudosphere.

6. There is an interesting piece of electronic music called "YLEM" by Karlheinz Stockhausen (Deutsche Grammophon recording DGG-2530442) that incorporates some of George Gamow's ideas about an oscillating Universe. The players begin in a tight circle on stage and then expand outward until they reach a certain point, from which they (and the music) return to the original circle. (In this connection, a piece called "The Creation" by Vincent Persichetti also makes reference to the big bang.)

7. It can be an interesting project to compare our theories of cosmology with those of ancient cultures; there are surprising similarities. See, for example:

   Blacker, C., and M. Loewe. *Ancient Cosmologies* (Rowman and Littlefield, 1975).
   Branley, F. "Conceptions of the Universe," *Natural History*, December 1970.
   Long, C. *Alpha--The Myths of Creation* (Collier paperback, 1963).

8. The text judiciously limits its discussion to the so-called "Establishment theories" in cosmology. For that rare instructor or student who has extra time at the end of an astronomy course, here are some references for investigating some of the more unusual theories.

   a. *The Hoyle-Narlikar cosmology:*
      Hoyle, F. *Astronomy and Cosmology: A Modern Course* (W. H. Freeman and Co., 1975).
      Hoyle, F. "The Future of Physics and Astronomy," *American Scientist*, March/April 1976.
      Kaufmann, W. "The Hoyle-Narlikar Cosmology," *Mercury*, May/June 1976.
      Thomsen, D. "Cosmology According to Hoyle," *Science News*, June 14, 1975.

   b. *The Lyttleton-Bondi theory of charge imbalance:*
      Bondi, H., et al. *Rival Theories of Cosmology* (Oxford Univ. Press, 1960).
      Coleman, J. *Modern Theories of the Universe* (Signet paperback, 1963). [See p. 58.]
      Kilmister, C. *The Nature of the Universe* (Dutton, 1971). [See pp. 187-189.]

   c. *Antimatter cosmology:*
      Alfvén, H. *Worlds and Antiworlds* (W. H. Freeman and Co., 1960).
      Alfvén, H. "Antimatter and Cosmology," *Scientific American*, April 1967.
      Thomsen, D. "Does the Universe Need Antimatter?" *Science News*, March 31, 1973.

9. For those who like their cosmological explanations fancifully done, there is a wonderful little book by George Gamow called *Tomkins in Paperback* (1940-45; reprinted as Cambridge Univ. Press paperback). The stories all concern Mr. Tomkins, a mild-mannered clerk who, while courting a physics professor's daughter, has some rather bizarre dreams about scientific subjects. Chapters 4 through 6 deal with curved space and cosmology.

   The Mr. Tomkins stories are mostly science with just a little fiction; there are some other interesting books, which are mainly fiction with a little science, that are appropriate for this chapter:

   Anderson, P. *Tau Zero* (British Coronet paperback, 1970). [In this amazing book, unfortunately out of print in the U.S.A., a spaceship traveling at very close to the speed of light, survives to see the Universe collapse. The relativity is quite accurate, the cosmology less so, but students who are interested in space travel really enjoy this book.]

Asimov, I. *The Gods Themselves* (Fawcett Crest paperback, 1972). [Dr. Asimov's award-winning science-fiction novel solves the origin of the big bang, among other things.]

Calvino, I. *Cosmicomics* (Collier paperback, 1965). [Short, moralistic allegories about the human condition, using cosmological themes.]

10. One interesting thought experiment that can be useful in considering the large-scale properties of the Universe is the paradox of Olbers and de Cheseaux, which asks, "If the Universe were infinite, how could the sky be dark at night?" See:

Harrison, E. "Why the Sky Is Dark at Night," *Physics Today*, February 1974.

Jaki, S. *The Paradox of Olbers' Paradox* (Herder & Herder, 1969).

11. For students who wish to delve more deeply into general relativity, but do not have the mathematical sophistication required by a regular text in the field, the following references may be useful.

Bergmann, P. *The Riddle of Gravitation* (Scribner's, 1968).

Frank, P. "Relativity and Its Astronomical Implications" (a *Sky and Telescope* reprint pamphlet available from Sky Publishing Corp.).

Kaufmann, W. *Relativity and Cosmology* (Harper & Row, 1976).

Kaufmann, W. *Cosmic Frontiers of General Relativity* (Little, Brown, 1977).

Klein, H. *The New Gravitation* (Lippincott, 1971).

12. The California poet Robinson Jeffers wrote eloquent and informed poetry using astronomical imagery, in part because his brother was an astronomer at Lick observatory. See, in particular, such poems as "Margrave" or "The Great Explosion," which are based on cosmological concepts.

READING MATERIALS FOR THE STUDENT

Historical Books

Charon, J. *Cosmology* (McGraw-Hill paperback, 1969).

Clayton, D. *The Dark Night Sky: A Personal Adventure in Cosmology* (Quadrangle Books, 1975).

Coleman, J. *Modern Theories of the Universe* (Signet paperback, 1963).

Coleman, J. *Early Theories of the Universe* (Signet paperback, 1967).

Dickson, F. *The Bowl of Night* (Philips Technical Library, 1968).

Gamow, G. *The Creation of the Universe*, 2nd ed. (Bantam paperback, 1961). [A classic little book by a noted cosmologist.]

Munitz, M. *Space, Time and Creation* (Collier paperback, 1957).

Munitz, M., (ed.). *Theories of the Universe* (Free Press paperback, 1957). [A nice set of nontechnical readings by the major figures in cosmology.]

Books on Modern Cosmology

Bonnor, W. *The Mystery of the Expanding Universe* (Macmillan, 1964). [Nice explanations; the observational sections are now dated.]
Calder, N. *The Violent Universe* (Viking paperback, 1969). [See Section III.]
Goldsmith, D., and D. Levy. *From the Black Hole to the Infinite Universe* (Holden-Day paperback, 1974).
Hodge, P. *Galaxies and Cosmology* (McGraw-Hill paperback, 1966).
Hodge, P. *Concepts of the Universe* (McGraw-Hill paperback, 1969).
Hoyle, F. *From Stonehenge to Modern Cosmology* (W. H. Freeman and Co., 1972).
John, L., (ed.). *Cosmology Now* (Taplinger, 1973).
Kaufmann, W. *Relativity and Cosmology* (Harper & Row paperback, 1973).
Kilmister, C. *The Nature of the Universe* (Dutton paperback, 1971).
Saslaw, W., and K. Jacobs (eds.). *The Emerging Universe* (Univ. Press of Virginia, 1972).
Schatzman, E. *The Structure of the Universe* (McGraw-Hill paperback, 1968).
Shipman, H. *Black Holes, Quasars and the Universe* (Houghton-Mifflin paperback, 1976). [A superb primer on cosmology and other topics.]
Whitrow, G. *The Structure and Evolution of the Universe* (Harper Torchbook, 1959). [Observational sections now dated.]

Articles

Albers, D. "The Meaning of Curved Space," *Mercury*, July/August 1975.
Alfvén, H. "Antimatter and Cosmology," *Scientific American*, April 1967.
Burbidge, G., and E. Burbidge. "Modern Riddles in Cosmology," in *The Heritage of Copernicus*, ed. J. Neyman (MIT Press, 1974).
Callahan, J. "The Curvature of Space in a Finite Universe," *Scientific American*, August 1976.
Fowler, W. "The Origin of the Elements," *Scientific American*, September 1956.
Gamow, G. "The Evolutionary Universe," *Scientific American*, September 1956.
Gott, J., et al. "Will the Universe Expand Forever?" *Scientific American*, March 1976.
Gunn, J. "Will the Universe Expand Forever?" *Mercury*, Nov./Dec. 1975.
Hack, M. "The Abundance of Helium in the Cosmos," *Sky and Telescope*, Sept.-Oct. 1972.
Hoyle, F. "The Steady-state Universe," *Scientific American*, September 1956.
"Is Gravity Weakening?" *Science News*, August 24-31, 1975.
Jastrow, R. "How Old Is the Universe?" *Natural History*, Aug./Sept. 1974.
Kaufmann, W. "The Hoyle-Narlikar Cosmology," *Mercury*, May/June 1976.
Krogdahl, W. "The Creation of the Universe," *Sky and Telescope*, March 1973.
Layzer, D. "The Arrow of Time," *Scientific American*, December 1975.
Margon, B. "The Missing Mass," *Mercury*, Jan./Feb. 1975.
McCrea, W. "Cosmology Today," *American Scientist*, Sept./Oct. 1970.
Morrison, D., and N. Morrison. "The Expansion of the Universe," *Mercury*, July/August 1975.
Morrison, D., and N. Morrison. "An Infinitely Expanding Universe," *Mercury*, Sept./Oct. 1975. [See also "Noted in the Current Journals" in the March/April and May/June 1975 issues.]
Pasachoff, J., and W. Fowler. "Deuterium in the Universe," *Scientific*

American, May 1974.
Peebles, P., and D. Wilkinson. "The Primeval Fireball," *Scientific American*, June 1967.
Rees, M. "Progress and Prospects in High Energy Astrophysics," *Proceedings of the National Academy of Science U.S.A.*, December 1975.
Sandage, A. "The Age of Creation," in *Science Year 1968* (World Book Encyclopaedia).
Sandage, A. "The Time Scale for Creation," *Astronomical Society of the Pacific Leaflets* #477 and #478 (March-April 1969).
Schramm, D. "The Age of the Elements," *Scientific American*, January 1974.
Schucking, E., and B. Biram. "From Genesis to Doomsday," in *Science Year 1967* (World Book Encyclopaedia).
Thomsen, D. "Dirac's New Field Theory: A Strange Variable Gauge Universe," *Science News*, February 9, 1974.
Thomsen, D. "A Cosmological Triple Play," *Science News*, February 16, 1974.
Thomsen, D. "The Weak Interaction in the Universe," *Science News*, May 31, 1975.
Thomsen, D. "Cosmology According to Hoyle," *Science News*, June 14, 1975.
Thomsen, D. "An Open or Shut Case," *Science News*, January 3, 1976.
Tinsley, B. "From Big Bang to Eternity," *Natural History*, October 1975.
van Flandern, T. "Is Gravity Getting Weaker?" *Scientific American*, February 1976.
Webster, A. "The Cosmic Background Radiation," *Scientific American*, August 1974.
Weymann, R. "Award of the Bruce Medal to Sir Martin Ryle," *Mercury*, Sept./Oct. 1974. [Discussion of radio-source counts.]
Wheeler, J. "Our Universe," *American Scholar*, Spring 1968.
Wheeler, J. "The Universe as Home for Man," *American Scientist*, Nov./Dec. 1974.
"Will the Universe Expand Forever?" *Science News*, November 8, 1975.

READING MATERIALS FOR THE INSTRUCTOR

Books

Bondi, H. *Cosmology* (Cambridge Univ. Press, 1961).
Longair, M., (ed.). *Confrontation of Cosmological Theories with Observational Data* (D. Reidel, 1974). [Symposium #63 of the International Astronomical Union.]
Misner, C., et al. *Gravitation* (W. H. Freeman and Co. paperback, 1973).
North, J. *The Measure of the Universe* (Oxford Univ. Press, 1965). [An historical/philosophical treatise.]
Peebles, P. *Physical Cosmology* (Princeton Univ. Press, 1971).
Rees, M., et al. *Black Holes, Gravitational Waves and Cosmology: An Introduction to Current Research* (Gordon and Breach, 1973).
Sciama, D. *Modern Cosmology* (Cambridge Univ. Press, 1971).

Articles

Alpher, R. "Large Numbers, Cosmology and Gamow," *American Scientist*, Jan./Feb. 1973.

Danziger, J. "The Cosmic Abundance of Helium," in *Annual Reviews of Astronomy and Astrophysics*, vol. 8 (1970).
Davis, M. "Galaxies and Cosmology," in *Frontiers of Astrophysics*, ed. E. Avrett (Harvard Univ. Press, 1976).
de Vaucouleurs, G. "The Case for a Hierarchical Cosmology," *Science*, February 27, 1970.
Ellis, G. "Cosmology and Verifiability," *Quarterly Journal of the Royal Astronomical Society*, September 1972.
Field, G. "Intergalactic Matter," in *Annual Reviews of Astronomy and Astrophysics*, vol. 10 (1972).
Field, G. "The Mass of the Universe: Intergalactic Matter," in *Frontiers of Astrophysics*, ed. E. Avrett (Harvard Univ. Press, 1976).
Gal-or, B. "The Crisis About the Origin of Irreversibility and Time Anisotropy," *Science*, April 7, 1972.
Harrison, E. "The Early Universe," *Physics Today*, June 1968.
Harrison, E. "The Cosmic Numbers," *Physics Today*, December 1972
Harrison, E. "Why the Sky Is Dark at Night," *Physics Today*, February 1974.
Harrison, E. "Standard Model of the Early Universe," in *Annual Reviews of Astronomy and Astrophysics*, vol. 13 (1975).
Jauncey, D. "Radio Surveys and Source Counts," in *Annual Reviews of Astronomy and Astrophysics*, vol. 13 (1975).
Narlikar, F. "The Interaction of Laboratory Science and Astronomy," *American Scientist*, May/June 1970.
Novikov, J., and Y. Zel'dovich. "Physical Processes Near Cosmological Singularities," in *Annual Reviews of Astronomy and Astrophysics*, vol. 11 (1973).
Partridge, R. "The Primeval Fireball Today," *American Scientist*, Spring 1969.
Rindler, W. "Relativistic Cosmology," *Physics Today*, November 1967.
Rothenberg, M. "Cosmic Background Radiation Reveals Its Blackbody Shape," *Physics Today*, July 1975. [In the "Search and Discovery" section.]
Sandage, A. "Cosmology: The Search for Two Numbers," *Physics Today*, February 1970.
Sandage, A. "Distance to Galaxies: The Hubble Constant, the Friedmann Time and the Edge of the World," *Quarterly Journal of the Royal Astronomical Society*, June 1972.
Schramm, D., and R. Wagoner. "What Can Deuterium Tell Us?" *Physics Today*, December 1974.
Thaddeus, P. "The Short-wavelength Spectrum of the Microwave Background," in *Annual Reviews of Astronomy and Astrophysics*, vol. 10 (1972).

AUDIO-VISUAL AIDS

Films

Flatland (1965, color, 12 min). This clever little animated film, made from E. Abbott's classic book, can be a nice introduction to thinking about non-Euclidean geometries. (Distributed by McGraw-Hill Films.)

The Immense Design (1960, black and white, 60 min). Fred Hoyle, William Fowler, Jesse Greenstein, and Allan Sandage discuss the various models of the Universe. Discussion is somewhat dated, but the stature of the participants makes this a significant film record. (NBC Television film; distributed by the California Institute of Technology.)

Powers of Ten (1968, color, 8 min). See the description in Chapter 1. A nice way to sum up the course.

Violent Universe (1969, black and white, 148 min). See description in Chapter 18.

Slides

Norton Scientific Company markets sets of slides on "Galaxies and Cosmology" and "General and Special Relativity." Some are a little exaggerated, but some are quite nice.

Audio Tapes

As It Was in the Beginning (26 min). Robert Jastrow discusses the evolution of the Universe. Similar to the epilogue of our text. (Available from the Center for Cassette Studies.)

Cosmology (30 min). Joseph Silk is interviewed by Richard Reis. (Available from the Astronomical Society of the Pacific.)

The Evolution in Theories of Gravitation (10 min). Robert Dicke discusses the various theories. (Available from Spring Green Multimedia.)

Space, Void and Cosmos: Hermann Bondi (30 min). A discussion about the origin of the Universe. (Available from the Center for Cassette Studies.)

SAMPLE EXAM QUESTIONS (MULTIPLE-CHOICE)

1. Astronomers estimate the age of the Universe to be about--
   a. 6 thousand years
   b. 4.5 billion years
   c. 20 billion years
   d. a thousand billion years
   e. There is not even a vague estimate about the age of the Universe.
2. The estimate of the age of the Universe made from considering the present-day recession of the galaxies needs to be modified to take into account the effects of--
   a. gravity
   b. the evolution of galaxies as the stars in them evolve
   c. the steady-state theory
   d. clustering in galaxies
   e. the fact that speeds were slower right after the big bang
3. The 3-degree background comes from--
   a. the combined radiation of all the distant galaxies
   b. clouds of hot material in clusters of galaxies
   c. radioactivity within the Earth and in the Van Allen belts
   d. the time when radiation and gas decoupled in the early Universe
   e. the time when the Universe was only a few seconds old
4. A hydrogen atom in the ocean was probably made--
   a. during the first few minutes of the Universe
   b. in and by stars
   d. Either of the above answers could be correct.
5. A helium atom in a balloon at a circus was probably made--
   a. during the first few minutes of the Universe

b. in and by stars
c. Either of the above answers could be correct.

6. A hydrogen atom in our Sun was probably made--
   a. during the first few minutes of the Universe
   b. in and by stars
   c. Either of the above answers could be correct.
7. A helium atom in our Sun was probably made--
   a. during the first few minutes of the Universe
   b. in and by stars
   c. Either of the above answers could be correct.
8. A uranium atom in the Earth's rocks was probably made--
   a. during the first few minutes of the Universe
   b. in and by stars
   c. Either of the above answers could be correct.
9. A carbon atom in the hand writing this exam was probably made--
   a. during the first few minutes of the Universe
   b. in and by stars
   c. Either of the above answers could be correct.
10. Which of the following is *not* a piece of evidence in favor of the big-bang picture of the beginning of the Universe?
   a. the age estimates for the oldest stars
   b. the discovery of heavier elements in the oldest stars
   c. the 3-degree background radiation
   d. the abundance of helium in the Universe
   e. the expansion of the galaxies
11. If the density of the Universe is larger than the critical density, then--
   a. the Universe is open and flat
   b. the Universe is open and negatively curved
   c. the sum of the angles of enormous triangles adds up to less than 180°
   d. the Universe is closed and negatively curved
   e. the Universe is closed and positively curved

Answers to Multiple-choice Questions

1. c 2. a 3. d 4. a 5. c 6. a 7. c 8. b 9. b 10. b 11. e

SAMPLE EXAM QUESTIONS (ESSAY)

1. Archbishop Ussher, several centuries ago, "calculated" that the Universe had been created in 4004 B.C. Discuss pieces of astronomical evidence that would contradict this proposal.
2. How do astronomers estimate the age of the Universe? Give at least two methods.

NOTE: Questions 1 and 2 could be combined to form a grand summary question on a final exam.

3. You're drinking a glass of water. Recount the early history of a typical hydrogen atom in the water. How and where were the atoms of the other elements in the water and the glass formed?
4. What evidence can you cite in favor of the big-bang hypothesis for the beginning of the Universe?
5. (a) What are the geometric rules that distinguish among flat, positively curved, and negatively curved universes? (b) How can we use these rules, together with astronomical observations, to

determine what sort of Universe we find ourselves in? (c) What are the problems with doing the observations?

6. What are the two possible fates of the Universe? Why can't we say today which is the actual fate our Universe will suffer?
7. You are a very rich industrialist, and before you die you'd like to know what the fate of the Universe will be. Describe what projects you would fund to give you the best chance of getting an answer to this problem in the next 25 years.
8. Why did Einstein introduce a "fudge factor" (the cosmological term) into his equations, and why is that factor not necessary today?

APPENDIX ONE

GENERAL READING MATERIALS FOR INTRODUCTORY ASTRONOMY

MAGAZINES THAT CARRY INTRODUCTORY-LEVEL ASTRONOMY ARTICLES

American Scientist. Some of the occasional astronomy articles are above the level of a beginning student, but some are just right.

Mercury (published by the Astronomical Society of the Pacific). Contains review articles, bibliographies, interviews, interdisciplinary articles, summaries of current research, etc. (Replaced *Astronomical Society of the Pacific Leaflets.*)

National Geographic. Good, superbly illustrated astronomy articles have been appearing about once or twice a year in the last few years; strong emphasis on the Solar System and the space program.

Natural History. Regular short articles on astronomy (about one or two per issue), often written by well-known astronomers.

Science News. Regular coverage of new discoveries or theoretical work; occasional longer articles on interesting current topics.

Scientific American. Features an astronomy article about once every two issues, generally by prominent astronomers.

Sky and Telescope (published by Sky Publishing Corp.). Aimed at the amateur; features observing hints, telescope descriptions, excellent photographs, articles on current topics, etc.

Smithsonian. Occasional excellent articles on astronomical subjects.

GOOD GENERAL ASTRONOMY BOOKS (FOR THE STUDENT)

Bergamini, D. *Universe* (Life Nature Library, 1966).

Bova, B. *The New Astronomies* (Mentor Books paperback, 1972).

Calder, N. *Violent Universe* (Viking paperback, 1969).

Goldsmith, D., and D. Levy. *From the Black Hole to the Infinite Universe* (Holden-Day paperback, 1974).

Hodge, P. *The Revolution in Astronomy* (Holiday House paperback, 1970).

Hoyle, F. *Astronomy* (Doubleday, 1962). An illustrated history.

Jastrow, R. *Red Giants and White Dwarfs* (Signet paperback, 1971).

Kilmister, C. *The Nature of the Universe* (Dutton paperback, 1971).

Kopal, Z. *Man and His Universe* (Morrow paperback, 1971).

Ronan, C. *Invisible Astronomy* (Lippincott, 1972).

Sagan, C. *The Cosmic Connection* (Dell paperback, 1973).

Shipman, H. *Black Holes, Quasars, and the Universe* (Houghton-Mifflin paperback, 1976).

GOOD GENERAL ASTRONOMY BOOKS (FOR THE INSTRUCTOR)

Harwit, M. *Astrophysical Concepts* (Wiley, 1973).

Hoyle, F. *Astronomy and Cosmology* (W. H. Freeman and Co., 1975).

Rose, W. *Astrophysics* (Holt, Rinehart and Winston, 1973).

Smith, E., and K. Jacobs. *Introductory Astronomy and Astrophysics* (Saunders, 1973).

Swihart, T. *Astrophysics and Stellar Astronomy* (Wiley, 1968).

Weekes, T. *High Energy Astrophysics* (Chapman and Hall, 1969).

COLLECTIONS OF READINGS

Avrett, E., (ed.). *Frontiers of Astrophysics* (Harvard Univ. Press, 1976). [Somewhat advanced essays.]

Cornell, J., and E. Hayes (eds.). *Man and Cosmos* (Norton, 1975). [From a 1972 Smithsonian lecture series.]

Gingerich, O., (ed.). *New Frontiers in Astronomy* (W. H. Freeman and Co., 1975). [Reprints from *Scientific American*. See also a 1970 volume titled *Frontiers in Astronomy*.]

(Harvard) Project Physics Reader, Volume 2: Motion in the Heavens (Holt, Rinehart and Winston, 1970). [Part of a highschool physics curriculum.]

Munitz, M., (ed.). *Theories of the Universe* (Free Press, 1957). [Nontechnical excerpts from the writings of great astronomers of the past.]

The New Universe (Rand McNally, 1967-68). [Reprints of articles from the British magazine *Science Journal*.]

Neyman, J., (ed.). *The Heritage of Copernicus: Theories Pleasing to the Mind* (MIT Press, 1974). [Some nice historical essays.]

Page, T., and L. Page (eds.). "Sky and Telescope Reprint Series" (Macmillan, 1965-69). Includes *Wanderers in the Sky*, *Neighbors of the Earth*, *The Origin of the Solar System*, *Telescopes*, *Starlight*, *The Evolution of the Stars*, *Stars and Clouds of the Milky Way*, and *Beyond the Milky Way*

Saslaw, W., and K. Jacobs (eds.). *The Emerging Universe* (Univ. Press of Virginia, 1972). [Some essays are advanced; some more simple.]

Young, L., (ed.). *Exploring the Universe*, 2nd ed. (Oxford Univ. Press, 1971). [Mainly historical and philosophical readings.]

BIBLIOGRAPHIES

Berendzen, R., and D. DeVorkin. "Resource Letter EMAA-1: Educational Materials in Astronomy and Astrophysics," *American Journal of Physics*, June 1973. [Reprints available from the American Association of Physics Teachers.]

Fraknoi, A. "Book Trek," a regular column of annotated reading lists in *Mercury*. [Reprints are available in sets from the Astronomical Society of the Pacific.]

Kemp. D. A. *Astronomy and Astrophysics: A Bibliographical Guide* (Archon Books, 1970). [A somewhat haphazardly selected 582-page annotated guide to astronomical articles classified by subject area.]

Pfendl, H. "Physics for Poets, Pianists and Philosophers," *The Physics*

Teacher, May 1972. [An excellent annotated bibliography of introductory and interdisciplinary physical science readings.]

Ryan, M., and L. Shepley. "Resource Letter RC-1: Cosmology," *American Journal of Physics*, March 1976. [Includes references on all levels.]

Science for Society. [An annotated bibliography of references prepared yearly by the A.A.A.S. Commission on Science Education, 1515 Massachusetts Avenue, N.W., Washington, D.C. 20005.]

APPENDIX TWO
USEFUL ADDRESSES

American Association for the Advancement of Science, 1515 Massachusetts Avenue, N.W., Washington, D.C. 20005. [Publishes *Science* magazine, organizes conferences, issues bibliographies, etc.]

American Association of Physics Teachers, Graduate Physics Building, S.U.N.Y. at Stony Brook, Stony Brook, N.Y. 11794. [Publishes the *American Journal of Physics* and *The Physics Teacher*, as well as many pamphlets about physics teaching; maintains a repository of films, loops, etc.; has regional chapters.]

American Association of Variable Star Observers, 187 Concord Avenue, Cambridge, Mass. 02138. [Organization of amateur and professional astronomers who observe variable phenomena.]

American Astronomical Society, (H. Gurin, Executive Officer), 211 FitzRandolph Road, Princeton, N.J. 08540. [The national organization for research astronomers; has career pamphlets, runs visiting lectureships program. See also "Task Group on Education in Astronomy," a committee of the A.A.S.]

American Institute of Physics, 335 East 45th Street, New York, N.Y. 10017. [Publishes *Physics Today*; coordinates the various physical science professional societies and their activities.]

Annual Reviews, Inc., 4139 El Camino Way, Palo Alto, Calif. 94306. [Publishes *Annual Review of Astronomy and Astrophysics* and other review books.]

Association of Lunar and Planetary Observers, Box 3AZ, University Park, N. Mex. 88001. [Amateur astronomy organization interested in Solar System studies.]

Astronomical League, 1719 N Street, N.W., Washington, D.C. 20036. [Umbrella organization for many amateur astronomy groups.]

Astronomical Society of the Pacific, 1244 Noriega Street, San Francisco, Calif. 94122. [Publishes *Mercury*; supplies slides, tapes, teaching aids, star maps, etc.]

Bailey Film Associates. See "BFA/Ealing."

BFA/Ealing, Educational Media Division, 2211 Michigan Avenue, Santa Monica, Calif. 90404. [Distributor of 8 mm and super-8 film loops.]

California Institute of Technology Bookstore, 1201 East California Boulevard, Pasadena, Calif. 91109. [Distributes Hale Observatories materials, films, and the *Hubble Atlas of Galaxies*.]

Celestron International, 2835 Columbia Street, Torrance, Calif. 90503. [Markets compact (Schmidt-Cassegrain) telescopes and accessories.]

Center for Cassette Studies, 8110 Webb Avenue, North Hollywood, Calif. 91605. [Distributes audio cassettes.]

Central Scientific Company, 2600 S. Kostner Avenue, Chicago, Ill. 60623. [Distributor of physics and related laboratory and demonstration equipment.]

Edmund Scientific Company, Edscorp Building, Barrington, N.J. 08007. [Markets inexpensive telescopes and many other science materials.]

Hale Observatories. [Prints and slides are available from California Institute of Technology Bookstore; see address above.]

Indiana University Audio-Visual Center, Indiana University, Bloomington, Ind. 47401. [Large distributor of science films.]

International Film Bureau, 332 S. Michigan Avenue, Chicago, Ill. 60604. [Film distributor.]

Lick Observatory, University of California, Santa Cruz, Calif. 95060. [Distributes prints and slides.]

McGraw-Hill Films, 330 West 42nd Street, New York, N.Y. 10018. [Distributor of many good science films.]

Modern Learning Aids, 1212 Avenue of the Americas, New York, N.Y. 10036. [Film distributor.]

National Academy of Science, Washington, D.C. 20418.

National Aeronautics and Space Administration (N.A.S.A.) [Distributes educational publications, information sheets on recent missions, and free films. Contact the office nearest you from the following list:]

- Educational Programs and Services Office, National Aeronautics and Space Administration, Washington, D.C. 20546.
- Educational Programs and Services, N.A.S.A. Goddard Space Flight Center, Greenbelt, Md. 20771.
- Educational Programs and Services, N.A.S.A. Langley Research Center, Hampton, Va. 23365.
- Community Development Office, N.A.S.A. John F. Kennedy Space Center, Florida 32899.
- Educational Liason Branch, N.A.S.A. Marshall Space Flight Center, Huntsville, Alabama 35812.
- Educational Programs and Services, N.A.S.A. Lewis Research Center, 21000 Brookpark Road, Cleveland, Ohio 44135.
- Educational Programs and Services, Public Affairs Office, N.A.S.A. Manned Spacecraft Center, Houston, Texas 77058.
- Educational Services, N.A.S.A. Ames Research Center, Moffett Field, Calif. 94035.

Norton Scientific Company, P.O. Box 4294, Tucson, Ariz. 85717. [Markets slides and planetarium equipment.]

Omnibooks, 900 Major Sherman Lane, Monterey, Calif. 93940. [An astronomical book discount service.]

Science Service, 1719 N Street, N.W., Washington, D.C. 20036. [Publishes *Science News* and science kits.]

Sky Publishing Corporation, 49-50-51 Bay State Road, Cambridge, Mass. 02138. [Publishes *Sky and Telescope*; distributor of atlases, books for amateur astronomers, lab exercises, etc.]

Spring Green Multimedia, P.O. Box 9015, Washington, D.C. 20003. [Distributes cassettes and other audio-visual materials.]

Task Group on Education in Astronomy, c/o Paul Knappenberger, Science Museum of Virginia, 217 Governor Street, Richmond, Va. 23219. [The education committee of the A.A.S.]

University of California Extension Media Center, University of California, Berkeley, Calif 94720. [Large distributor of science films.]

University of Illinois Visual Aids Service, University of Illinois, 1325 South Oak Street, Champaign, Ill. 61820. [Large distributor of science films.]

Yerkes Observatory, The Photographic Service Dept., Williams Bay, Wisconsin 53191. [Distributes prints, slides, and transparencies.]

APPENDIX THREE
LABORATORY AND OBSERVATIONAL EXERCISES AND MANUALS

1. A set of six laboratory exercises edited by Owen Gingerich is available from Sky Publishing Corporation. They also market the ESSCO series of charts, maps, and grids. Write for catalog. (See address in Appendix 2.)

2. *Mercury* magazine (published by the Astronomical Society of the Pacific) has carried a number of lab exercises as part of their semiregular feature on education, "Universe in the Classroom."

3. L.S.W. Associates (P.O. Box 82, Mattituck, N.Y. 11952) publishes a series of pamphlets, "Astronomy Through Practical Investigations." They will send complimentary samples upon request.

4. Paul Knappenberger has compiled a pilot set of elementary activities in astronomy for the highschool level. A few can be adapted to college courses, and further such sets may be gathered in the future. (See address for Task Group on Education in Astronomy, Appendix 2.)

5. The following lab manuals are available from commercial publishers.

   Atanasijevik, I. *Selected Exercises in Galactic Astronomy* (Springer-Verlag, 1971). [More advanced.]
   Culver, R. *An Introduction to Experimental Astronomy* (W. H. Freeman and Co., 1974).
   Gainer, M. *Astronomy: Observational Activities and Experiments* (Allyn and Bacon, 1974).
   Holzinger, J., and M. Seeds. *Laboratory Exercises in Astronomy* (Macmillan, 1976).
   Huffer, C., and R. Marasso. *Laboratory Exercises in Introductory Astronomy* (Holt, Rinehart and Winston, 1967).
   Kelsey, L., D. Hoff, and J. Neff. *Astronomy: Activities and Experiments* (Kendall/Hunt, 1974).
   Minnaert, M. *Practical Work in Elementary Astronomy* (Springer-Verlag, 1969).
   Shaw, R., and S. Boothroyd. *Manual of Astronomy* (W. Brown, 1967).

6. A long annotated list, "Laboratory Experiences for Elementary Astronomy," by H. Kruglak, appeared in the *American Journal of Physics*, September 1976. Reprints are available from the American Association of Physics Teachers.

7. Planetarium references:

a. *The Planetarium and Atmospherium*, by O. Norton (Naturegraph Publishers, Healdsburgh, California 95448; 1968), is a nice introduction to the history and use of the planetarium.

b. The International Society of Planetarium Educators (I.S.P.E.) publishes a regular journal, *The Planetarian*. Write to William Fagan, Oakton High School, 2900 Sutton Road, Vienna, Va. 22180.

c. Spitz Space Systems (Chadds Ford, Pa. 19317), one of the major planetarium manufacturers, makes available a number of publications about the use of planetaria as instructional tools. They also run workshops for teachers.

8. A brief list of handbooks for observing:

Brown, S. *All About Telescopes* (available from Sky Publishing Corp. or Edmund Scientific Co.).

Jones, K. *Messier's Nebulae and Star Clusters* (American Elsevier, 1968). [An exhaustive observing guide to extended objects.]

Menzel, D. *Field Guide to Stars and Planets* (Houghton-Mifflin, 1964).

Muirden, J. *The Amateur Astronomer's Handbook* (T. Crowell, 1974).

Neely, H. *A Primer for Stargazers* (Harper & Row, 1970). [Good for beginners who want only the "where" and not the "why."]

Norton, A. *Norton's Star Atlas and Reference Handbook*, 16th ed. (Sky Publishing Corp., 1973).

Paul, H. *Telescope for Skygazing*, and *Outer Space Photography for the Amateur* (both available from Sky Publishing Corp.).

Peltier, L. *Guideposts to the Stars* (Macmillan, 1972). [Very elementary; good for beginners.]

Thompson, A. *Making Your Own Telescope* (available from Sky Publishing Corp.).

Webb, T., and W. Mayall. *Celestial Objects for Common Telescopes*, 2 vols. (Dover paperbacks, 1962).

APPENDIX FOUR
INDEX TO SUBJECTS COVERED IN "TOPICS FOR DISCUSSION" SECTIONS

The number before the decimal point indicates the chapter; the number after the decimal point indicates the topic number.